Lecture Notes in Mathematics

Edited by A. Dold and B. Eckmann

304

A. K. Bousfield
D. M. Kan

Homotopy Limits, Completions and Localizations

Springer-Verlag
Berlin Heidelberg New York London Paris Tokyo

Authors

Aldridge K. Bousfield
University of Illinois, Department of Mathematics
Chicago, Illinois 60680, USA

Daniel M. Kan
Massachusetts Institute of Technology
Cambridge, MA 02139, USA

1st Edition 1972
2nd corrected Printing 1987

Mathematics Subject Classification (1970): 18A30, 18A35, 55-02, 55D05, 55D10, 55D99

ISBN 3-540-06105-3 Springer-Verlag Berlin Heidelberg New York
ISBN 0-387-06105-3 Springer-Verlag New York Berlin Heidelberg

2146/3140-543210

Contents

Part I. Completions and localizations

§0. Introduction to Part I

Our main purpose in Part I of these notes, i.e. Chapters I through VII, is to develop for a ring R a functorial notion of R-completion of a space X which

(i) for $R = Z_p$ (the integers modulo a prime p) and X subject to the usual finiteness conditions, coincides, up to homotopy, with the p-profinite completions of [Quillen (PG)] and [Sullivan, Ch.3], and

(ii) for $R \subset Q$ (i.e. R a subring of the rationals), coincides, up to homotopy, with the localizations of [Quillen (RH)], [Sullivan, Ch.2], [Mimura-Nishida-Toda] and others.

Our R-completion is defined for arbitrary spaces, and throughout these notes we have tried to avoid unnecessary finiteness and simple connectivity assumptions. To develop our R-completion we need some homotopy theoretic results on towers of fibrations, cosimplicial spaces, and homotopy limits, which seem to be of interest in themselves and which we have therefore collected in Part II of these notes, i.e. Chapters VIII through XII.

There are, we believe, two main uses for completions and localizations, i.e. for R-completions: first of all, they permit a "fracturing of ordinary homotopy theory into mod-p components"; and secondly, they can be used to construct important new (and old) spaces.

Of course, the general idea of "fracturing" in homotopy theory is very old; and indeed, the habit of working mod-p or using Serre's

$\mathcal{C}$-theory is deeply ingrained in most algebraic topologists. However "fracturing" in its present form (due largely to Sullivan) goes considerably further and, among other things, helps explain the efficacy of the familiar mod-p methods. Roughly speaking (following Sullivan), one can use completions or localizations to "fracture" a homotopy type into "mod-p components" together with coherence information over the rationals; and the original homotopy type can then be recovered by using the coherence information to reassemble the "mod-p components". In practice the rational information often "takes care of itself", and ordinary homotopy theoretic problems (e.g. whether two maps are homotopic or whether a space admits an H-space structure) often reduce to "mod-p problems". Of course, the "world of mod-p homotopy" is interesting in its own right (e.g. see [Adams (S)]).

As remarked above, another use for R-completions is to construct important spaces. It is, in fact, now standard procedure to use localization methods, e.g. Zabrodsky mixing, to construct new finite H-spaces. As other examples, we note that the space $(\Omega^{\infty}S^{\infty})_{(0)}$ is homotopy equivalent to the Z-completion of $K(S_{\infty},1)$, where S_{∞} is the "infinite symmetric group"(see Ch.VII, 3.4), and that, for the Z_p-completion of certain spheres, one can obtain classifying spaces by Z_p-completing suitable non-simply connected spaces (see [Sullivan] and Ch.VII, 3.6). Examples of this sort also seem to be useful in (higher dimensional) algebraic K-theory.

Some more comments are required on the relation between our R-completion and the completions and localizations of others:

In the case $R \subset Q$, as previously noted, our R-completion agrees, up to homotopy, with the <u>localizations</u> proposed by other authors; essentially, we have generalized the localization to non-simply connected spaces.

The situation for $R = Z_p$ is more complicated. Two homotopically equivalent versions of the p-profinite completion have been proposed by [Quillen (PG)] and [Sullivan, Ch.3] for arbitrary spaces; and it can be shown that our Z_p-completion and their p-profinite completion do not coincide, up to homotopy, for arbitrary spaces, although they do for spaces with Z_p-homology of finite type. One difficulty with the p-profinite completion is that for many simply connected spaces (e.g. for $K(M,n)$ where M is an infinite dimensional Z_p-module) the iterated p-profinite completion is not homotopy equivalent to the single one. This difficulty is avoided by the Z_p-completion. Nevertheless, the p-profinite completion remains very interesting, even when it differs from the Z_p-completion.

Some further general advantages of the R-completion are worth mentioning:

(i) Up to homotopy, the R-completion preserves fibrations under very general conditions (namely, when the fundamental group of the base acts "nilpotently" on the R-homology of the fibre).

(ii) Very many spaces X are R-good, i.e. the canonical map from X to its R-completion preserves R-homology and is, up to homotopy, "terminal" among the maps with this property; for instance, if $R \subset Q$ or $R = Z_p$, then all simply connected spaces are R-good, and so are many others (see Chapters V, VI and VII).

(iii) The mod-R homotopy spectral sequence of [Bousfield-Kan (HS)] can be used to relate the R-homology of a space with the homotopy groups of its R-completion.

(iv) The R-completion of a $K(\pi,1)$ has interesting group theoretic significance. For example, the Malcev completion of a nilpotent group π can be obtained as the fundamental group of the Q-completion of $K(\pi,1)$, a fact that suggests how to obtain "Malcev completions with respect to subrings of the rationals" (see Chapter V).

Similarly, the homotopy groups of the Z_p-completions of such a $K(\pi,1)$ have group theoretic significance (see Chapter VI).

Part I of these notes consists of seven chapters, the first four of which deal with the general theory, while the other three are concerned with various applications for $R \subset Q$ and $R = Z_p$. In more detail:

Chapter I. The R-completion of a space. Here we define the R-completion, $R_\infty X$, of a space X, and prove some of its basic properties, such as, for instance, the key property:

(i) A map $X \to Y$ induces an isomorphism on reduced R-homology

$$\tilde{H}_*(X; R) \approx \tilde{H}_*(Y; R)$$

if and only if it induces a homotopy equivalence between the R-completions

$$R_\infty X \simeq R_\infty Y.$$

Other (not very surprising) properties are:

(ii) The n-type of $R_\infty X$ depends only on the n-type of X.

(iii) Up to homotopy, the R-completion commutes with arbitrary disjoint unions and with finite products, and preserves multiplicative structures.

(iv) There is a generalization to a (functorial) fibre-wise R-completion.

We define $R_\infty X$ by first constructing a cosimplicial diagram of spaces $\underset{\sim}{R}X$, next associating with this a tower of fibrations $\{R_s X\}$, and finally defining the R-completion of X as the inverse limit

$R_\infty X$ of the tower $\{R_s X\}$. Justifications for this definition will be given in Chapters III and XI, where we show that $R_\infty X$ can, in two different ways, be considered as an "Artin-Mazur-like R-completion of X".

A useful tool in handling the R-completion is the homotopy spectral sequence of the tower of fibrations $\{R_s X\}$. This turns out to be the same as the homotopy spectral sequence $\{E_r(X; R)\}$ of X with coefficients in R of Bousfield-Kan, which, for $R = Z_p$, is the unstable Adams spectral sequence, while, for $R = Q$, this spectral sequence consists of the primitive elements in the rational cobar spectral sequence.

At the end of Chapter I we discuss the role of the ring R and show that, for all practical purposes, one can restrict oneself to the rings $R = Z_p$ (p prime) and $R \subset Q$.

Chapter II. Fibre lemmas. For a general fibration of connected spaces $F \to E \to B$, the map $R_\infty E \to R_\infty B$ is always a fibration, but its fibre need not have the same homotopy type as $R_\infty F$. However, there is a mod-R fibre lemma, which states that, up to homotopy, the R-completion preserves fibrations of connected spaces $F \to E \to B$, for which "$\pi_1 B$ acts nilpotently on each $\tilde{H}_i(F; R)$". This condition is, for instance, satisfied if the fibration is principal, or if B is simply connected.

This fibre lemma is a very useful result. It will, for instance, be used in the Chapters V and VI, to compute $\pi_* RX$ in terms of $\pi_* X$, for nilpotent X (i.e. connected X for which, up to homotopy, the Postnikov tower can be refined to a tower of principal fibrations).

Chapter III. Tower lemmas. A convenient feature of our definition of R-completion is its functoriallity. Still, it is often useful to have a more flexible (i.e. up to homotopy) approach available and we therefore prove in this chapter various tower lemmas, which give rather simple sufficient conditions on a tower of fibrations $\{Y_s\}$, in order that it can be used to obtain the homotopy type of the R-completion of a given space X. The strongest of these is the R-nilpotent tower lemma which states roughly:

If $\{Y_s\}$ is a tower of fibrations, together with compatible maps $X \to Y_s$, such that

(i) for every R-module M

$$\varinjlim H^*(Y_s; M) \approx H^*(X; M)$$

(ii) each Y_s is R-nilpotent (i.e. its Postnikov tower can, up to homotopy, be refined to a tower of principal fibrations with simplicial R-modules as fibres),

then, in a certain precise sense, the tower $\{Y_s\}$ has the same homotopy type as the tower $\{R_s X\}$ and hence the inverse limit spaces

$$R_\infty X = \varprojlim R_s X \quad \text{and} \quad \varprojlim Y_s$$

have the same homotopy type.

We also observe that $R_\infty X$ is an Artin-Mazur-like R-completion of X, as the results of this chapter imply that, up to homotopy, the tower $\{R_s X\}$ is cofinal in the system of R-nilpotent target spaces of X.

Chapter IV. An R-completion of groups and its relation to the R-completion of spaces. Here we use the greater flexibility of Chapter III, to obtain a more group-theoretic approach to the R-completion. For this we first define an Artin-Mazur-like R-completion of groups, which, for finitely generated groups and $R = Z_p$, reduces to the p-profinite completion of Serre, and which, for nilpotent groups and $R = Q$, coincides with the Malcev completion. Like any functor on groups, this R-completion functor from groups to groups can be "prolonged" to a functor from spaces to spaces, and we show that the latter is homotopically equivalent to the functor R_∞.

As an application we give a very short proof of Curtis' fundamental convergence theorem for the lower central series spectral sequence, at the same time extending it to nilpotent spaces.

Chapter V. Localizations of nilpotent spaces. The main purpose of this chapter is, to show that, for $R \subset Q$, the R-completion of a nilpotent space (i.e. a space for which, up to homotopy, the Postnikov tower can be refined to a tower of principal fibrations) is a localization with respect to a set of primes, and that therefore various well-known results about localizations of simply connected spaces remain valid for nilpotent spaces.

As an illustration we discuss some fracture lemmas (i.e. lemmas which state that, under suitable conditions, a homotopy classification problem can be split into a "rational problem" and "problems involving various primes or sets of primes") and their application to H-spaces.

We also prove that the homotopy spectral sequence $\{E_r(X; R)\}$ converges strongly to $\pi_* R_\infty X$ for $R \subset Q$ and X nilpotent.

Chapter VI. p-completions of nilpotent spaces. This chapter parallels Chapter V: We discuss the p-completion, i.e. the "up to homotopy" version of the Z_p-completion, for nilpotent spaces. This p-completion is merely a generalization of the familiar p-profinite completion for simply connected spaces of finite type, and we prove that various well-known results for such p-profinite completions remain valid for p-completions of nilpotent spaces.

As an illustration we discuss an arithmetic square fracture lemma, which states that, under suitable conditions, a homotopy classification problem can be split into "Z_p-problems" and a "rational problem".

We also obtain convergence results for the homotopy spectral sequence $\{E_r(X; Z_p)\}$ of a nilpotent space X, and observe that the same arguments apply to the lower p-central series spectral sequences.

Chapter VII. A glimpse at the R-completion of non-nilpotent spaces. It is clear from the results of Chapters V and VI that, for nilpotent spaces, the R-completion is quite well understood; however, very little is known about the R-completion of non-nilpotent spaces. In this last chapter of Part I we therefore discuss some examples of non-nilpotent spaces which indicate how much more work remains to be done.

We also make, at the end of this chapter, some comments on possible R-homotopy theories, for $R \subset Q$ and $R = Z_p$.

Warning!!! These notes are written simplicially, i.e. whenever we say

space we mean simplicial set.

However, in order to help make these notes accessible to a reader who knows homotopy theory, but who is not too familiar with simplicial techniques, we will in Chapter VIII, i.e. at the beginning of Part II:

(i) review some of the basic notions of simplicial homotopy theory, and

(ii) try to convince the reader that this simplicial homotopy theory is equivalent to the usual topological homotopy theory. Moreover, we have, throughout these notes, tried to provide the reader with references, whenever we use simplicial results or techniques, which are not an immediate consequence of their well-known topological analogues.

Some of the results of Part I of these notes were announced in [Bousfield-Kan (HR) and (LC)].

In writing Part I we have been especially influenced by the work of Artin-Mazur, Emmanuel Dror, Dan Quillen and Dennis Sullivan.

Chapter I. The R-completion of a space

§1. Introduction

In this chapter we define, for every space X and (commutative) ring R, a functorial R-completion of X and prove some of its basic properties. We also show that there is a corresponding notion of fibre-wise R-completion. In more detail:

§2, §3 and §4 Here we define the R-completion of X by first constructing a cosimplicial diagram of spaces $\underset{\sim}{R}X$, next associating with this a tower of fibrations $\{R_s X\}$, and finally defining the R-completion of X as the inverse limit $R_\infty X$ of the tower $\{R_s X\}$. It turns out that this R-completion comes with a natural map

$$\phi\colon\ X \longrightarrow R_\infty X.$$

Justifications for this definition will be given in Chapters III and XI, where we show that, up to homotopy, "$R_\infty X$ is an Artin-Mazur-like R-completion of X" in two different ways.

An immediate consequence of this definition is the existence of the associated spectral sequence, i.e. the homotopy spectral sequence of the tower of fibrations $\{R_s X\}$, which is an important tool in handling the R-completion. This spectral sequence is nothing but the homotopy spectral sequence $\{E_r(X;R)\}$ of X with coefficients in R of [Bousfield-Kan (HS)], which for $R = Z_p$ (the integers modulo a prime p) is the unstable Adams spectral sequence, while for $R = Q$ (the rationals) this spectral sequence consists of the primitive elements in the rational cobar spectral sequence.

§5 Our main results here are:

(i) a map $f: X \to Y$ induces an isomorphism

$$\tilde{H}_*(X;R) \approx \tilde{H}_*(Y;R)$$

if and only if it induces a homotopy equivalence

$$R_\infty X \simeq R_\infty Y .$$

(ii) a space X is either "R-good" or (very) "R-bad", i.e. either the map $\phi: X \to R_\infty X$ induces an isomorphism $\tilde{H}_*(X;R) \approx \tilde{H}_*(R_\infty X;R)$ and the maps $\phi: R_\infty^k X \to R_\infty^{k+1} X$ are homotopy equivalences for all $k \geq 1$, or the induced map $\tilde{H}_*(X;R) \to \tilde{H}_*(R_\infty X;R)$ is not an isomorphism and none of the maps $\phi: R_\infty^k X \to R_\infty^{k+1} X$ $(k \geq 1)$ is a homotopy equivalence.

In Chapters V, VI and VII we give various examples of R-good spaces and we show there that "most" (but not all) spaces are R-good for $R \subset Q$ and $R = Z_p$. An example of a space which is Z_p-bad is an infinite wedge of circles (Ch. IV, 5.4).

§6 and §7 contain the useful, but not very surprising results that

(i) the homotopy type of $R_\infty X$ in dimensions $< k$ depends only on the homotopy type of X in dimensions $\leq k$ (this will be somewhat strengthened in Ch. IV, 5.1), and

(ii) up to homotopy, the R-completion functor commutes with (disjoint) unions and finite products and preserves multiplicative structures.

§8 contains the observation that the notion of R-completion can be generalized to a notion of fibre-wise R-completion, i.e. one can, for a fibration $X \to B$, construct in a functorial manner a fibration $\dot{R}_\infty X \to B$ of which the fibres are the R-completions of the fibres of the map $X \to B$.

§9 We end this chapter with an investigation of the role of the ring R and show that, for "most" rings R, the homotopy type of $R_\infty X$ is completely determined by the homotopy types of the completions of X with respect to the rings Z_p (p prime) and the subrings of the rationals Q.

Notation and terminology. We remind the reader that these notes are written simplicially, i.e.

space = simplicial set.

In particular in this chapter we will mainly work in the category $\mathcal{S}$ of spaces (i.e. simplicial sets). For more details on this category (and its relationship to the category $\mathcal{T}$ of topological spaces) see Chapters VIII, IX and X.

§2. The triple $\{R,\phi,\psi\}$ on the category of spaces

In preparation for the definition (in §4) of the completion of a space with respect to a ring R we consider here a functor

$$R: \mathcal{S} \longrightarrow \mathcal{S}$$

on the category of spaces and two natural transformations

$$\phi: \mathrm{Id} \longrightarrow R \qquad \text{and} \qquad \psi: R^2 \longrightarrow R$$

which have the properties:

(i) $\{R,\phi,\psi\}$ is a triple, i.e. [Eilenberg-Moore]

$$(R\phi)\phi = (\phi R)\phi \qquad \psi(R\psi) = \psi(\psi R) \qquad \psi(R\phi) = \mathrm{id} = \psi(\phi R)$$

(ii) For every choice of base point $* \in X$, there is a canonical isomorphism

$$\pi_* RX \approx \tilde{H}_*(X;R)$$

such that the composition

$$\pi_* X \xrightarrow{\pi_*\phi} \pi_* RX \approx \tilde{H}_*(X;R)$$

is the Hurewicz homomorphism [May, p. 50], and

(iii) A map $f: X \to Y \in \mathcal{S}$ induces an isomorphism

$$\tilde{H}_*(X;R) \approx \tilde{H}_*(Y;R)$$

if and only if it induces a homotopy equivalence

$$RX \simeq RY \in \mathcal{S} .$$

2.1 Definition of the triple. For a space (i.e. simplicial set) X and a commutative ring R (with unit), let $R \otimes X$ denote the simplicial R-module freely generated by the simplices of X (i.e. $(R \otimes X)_n$ is the free R-module on X_n) and let

$$\phi: X \longrightarrow R \otimes X \qquad \text{and} \qquad \psi: R \otimes (R \otimes X) \longrightarrow R \otimes X$$

respectively be the map given by $\phi x = 1x$ for all $x \in X$ and the R-module homomorphism given by $\psi(1y) = y$ for all $y \in R \otimes X$. Then we define RX as the subspace

$$RX \subset R \otimes X$$

consisting of the simplices

$$\Sigma\, r_i x_i \qquad \text{with} \quad \Sigma\, r_i = 1.$$

If $R^n X = R \cdots RX$ for $n > 1$, then one readily sees that the maps ϕ and ψ induce natural transformations

$$\phi: \mathrm{Id} \longrightarrow R \qquad \text{and} \qquad \psi: R^2 \longrightarrow R$$

and that $\{R, \phi, \psi\}$ is a triple on the category $\mathcal{J}$.

When one uses this triple it is often convenient to work in

2.2 A pointed situation. The simplicial set RX defined above does not inherit an R-module structure from the simplicial R-module $R \otimes X$, but only a kind of affine R-structure, which turns into an R-module structure the moment one chooses a base point. More precisely, if one chooses a base point $* \in X$ and denotes also by $* \subset X$ the subspace generated by it, then the composition

$$RX \xrightarrow{\text{incl.}} R \otimes X \xrightarrow{\text{proj.}} R \otimes X \,/\, R \otimes *$$

obviously is an isomorphism of simplicial sets. Thus, given a base point $* \in X$, <u>one can consider RX as a simplicial R-module</u>. Moreover, if one does this, then <u>the map $\psi\colon R^2X \to RX$ becomes an R-module homomorphism</u>.

This can be used to relate the functor $R\colon \mathcal{S} \to \mathcal{S}$ to

<u>2.3 The reduced homology functor</u> $\tilde{H}_*(\ ;R)$. The <u>reduced homology</u> of a <u>pointed</u> space X with coefficients in R can be defined by [May, p. 94]

$$\tilde{H}_*(X;R) = \pi_*(R\otimes X \,/\, R\otimes *).$$

Thus, <u>for every $X \in \mathcal{S}$ and choice of base point $* \in X$, the isomorphism $RX \approx R\otimes X \,/\, R\otimes *$ of 2.2 induces an isomorphism</u>

$$\pi_*RX \approx \tilde{H}_*(X;R)$$

Note that the reduced homology does <u>not really</u> depend on the base point; in fact we could equivalently have defined

$$\begin{aligned}\tilde{H}_*(X;R) &= \pi_* \ker(R\otimes X \longrightarrow R\otimes *)\\ &= \ker(\pi_*(R\otimes X) \longrightarrow \pi_*(R\otimes *)).\end{aligned}$$

The remaining properties of $\{R,\phi,\psi\}$, stated at the beginning of this §, are now readily verified.

We end with a

<u>2.4 Remark on the affine R-structure of RX</u>. If $y_1,\cdots,y_k \in RX_n$, $r_1,\cdots,r_k \in R$ and $\Sigma\, r_i = 1$, then the linear

combination $\Sigma r_i y_i$ is a well defined element of RX_n and does <u>not</u> depend on a choice of base point.

§3. The total space of a cosimplicial space

In §4 we will define the R-completion of a space X as "the total space of a cosimplicial space $\underset{\sim}{R}X$". We therefore recall here first the notions of cosimplicial space and total space of a cosimplicial space. For a more detailed discussion of these notions we refer the reader to Chapter X, §2 and §3.

3.1 Cosimplicial spaces. A cosimplicial space $\underset{\sim}{X}$ is a cosimplicial object over the category $\mathcal{S}$ of spaces, i.e. $\underset{\sim}{X}$ consists of

(i) for every integer $n \geq 0$ a space $\underset{\sim}{X}^n \in \mathcal{S}$, and

(ii) for every pair of integers (i,n) with $0 \leq i \leq n$ coface and codegeneracy maps

$$d^i\colon \underset{\sim}{X}^{n-1} \longrightarrow \underset{\sim}{X}^n \quad \text{and} \quad s^i\colon \underset{\sim}{X}^{n+1} \longrightarrow \underset{\sim}{X}^n \qquad \in \mathcal{S}$$

satisfying the cosimplicial identities of Chapter X, §2 (which are dual to the simplicial identities).

Similarly a cosimplicial map $f\colon \underset{\sim}{X} \to \underset{\sim}{Y}$ consists of maps

$$f\colon \underset{\sim}{X}^n \longrightarrow \underset{\sim}{Y}^n \qquad \in \mathcal{S}$$

which commute with the coface and codegeneracy maps.

An important example is

3.2 The cosimplicial standard simplex. This is the cosimplicial space $\underset{\sim}{\Delta}$ which in codimension n consists of the standard n-simplex $\Delta[n] \in \mathcal{S}$ and for which the coface and codegeneracy maps are the standard maps between them (Ch. X, 2.2 and Ch. VIII, 2.9 and

2.11).

Using this one can now define

3.3 The total space of a cosimplicial space. For a cosimplicial space $\underset{\sim}{X}$ its *total space* $\operatorname{Tot} \underset{\sim}{X}$ or $\operatorname{Tot}_\infty \underset{\sim}{X}$ is the function space

$$\operatorname{Tot} \underset{\sim}{X} = \hom(\underset{\sim}{\Delta},\underset{\sim}{X}) \qquad \varepsilon\ \mathcal{S}$$

i.e. the space which has as q-simplices the cosimplicial maps

$$\Delta[q] \times \underset{\sim}{\Delta} \longrightarrow \underset{\sim}{X} \quad .$$

Often it is useful to consider

3.4 The total space as an inverse limit. Let

$$\underset{\sim}{\Delta}^{[s]} \subset \underset{\sim}{\Delta} \qquad\qquad -1 \le s$$

denote the *simplicial s-skeleton* of $\underset{\sim}{\Delta}$, i.e. $\underset{\sim}{\Delta}^{[s]}$ consists in codimension n of the s-skeleton (Ch. VIII, 2.13) of $\Delta[n]$. Then one can form the function spaces

$$\operatorname{Tot}_s \underset{\sim}{X} = \hom(\underset{\sim}{\Delta}^{[s]}, \underset{\sim}{X}) \qquad \varepsilon\ \mathcal{S}$$

and the maps

$$\operatorname{Tot}_s \underset{\sim}{X} \longrightarrow \operatorname{Tot}_{s-1} \underset{\sim}{X} \qquad \varepsilon\ \mathcal{S}$$

induced by the inclusions $\underset{\sim}{\Delta}^{[s-1]} \subset \underset{\sim}{\Delta}^{[s]}$ and observe that

(i) $\operatorname{Tot}_{-1} \underset{\sim}{X} = *$

(ii) $\operatorname{Tot}_0 \underset{\sim}{X} \simeq \underset{\sim}{X}^0$

(iii) $\operatorname{Tot} \underset{\sim}{X} = \varprojlim \operatorname{Tot}_s \underset{\sim}{X}$.

We end with a comment on

<u>3.5 The augmented case</u>. If $\underset{\sim}{X}$ is <u>augmented</u>, i.e. comes with an <u>augmentation map</u> $d^0: \underset{\sim}{X}^{-1} \to \underset{\sim}{X}^0$ such that

$$d^0 d^0 = d^1 d^0 : \underset{\sim}{X}^{-1} \longrightarrow \underset{\sim}{X}^1 \qquad \in \mathcal{C}$$

then this augmentation map clearly induces maps

$$\underset{\sim}{X}^{-1} \longrightarrow \mathrm{Tot}_s\, \underset{\sim}{X} \qquad -1 \leq s \leq \infty$$

which are compatible with the maps $\mathrm{Tot}_s\, \underset{\sim}{X} \to \mathrm{Tot}_{s-1}\, \underset{\sim}{X}$.

§4. The R-completion of a space

In this section we

(i) use the triple $\{R,\phi,\psi\}$ of §2 to construct, for every space X, a cosimplicial space $\underset{\sim}{R}X$, its cosimplicial resolution, and then define the R-completion of X as the total space (§3) of this cosimplicial space $\underset{\sim}{R}X$, and

(ii) observe that this R-completion of X is the inverse limit of a tower of fibrations and that thus there is an associated homotopy spectral sequence.

We also mention the fact (to be proven in §9) that it is no restriction to assume that the ring R is solid, i.e. that the multiplication map $R \otimes_Z R \to R$ is an isomorphism. The most important examples of such rings are the rings $R = Z_p$ (the integers modulo a prime p) and $R \subset Q$ (the subrings of the rationals).

We start with describing

4.1 The cosimplicial resolution. Let R be a commutative ring (with unit) and let $X \in \mathcal{S}$. The cosimplicial resolution of X with respect to R then is the augmented (3.5) cosimplicial space $\underset{\sim}{R}X$ given by

$$(\underset{\sim}{R}X)^k = R^{k+1} X$$

in codimension k and

$$((\underset{\sim}{R}X)^{k-1} \xrightarrow{d^i} (\underset{\sim}{R}X)^k) = (R^k X \xrightarrow{R^i\phi R^{k-i}} R^{k+1} X)$$

$$((\underset{\sim}{R}X)^{k+1} \xrightarrow{s^i} (\underset{\sim}{R}X)^k) = (R^{k+2} X \xrightarrow{R^i\psi R^{k-i}} R^{k+1} X)$$

as coface and codegeneracy maps.

Now we are ready for the definition of

4.2 The R-completion of a space. For $X \in \mathcal{S}$, its R-completion will be the total space (3.3)

$$R_\infty X = \text{Tot } \underset{\sim}{R}X \qquad \in \mathcal{S}$$

and, as $\underset{\sim}{R}X$ is augmented, this R-completion comes with a natural map (3.5)

$$\phi: X \longrightarrow R_\infty X \qquad \in \mathcal{S} .$$

Justification for this definition will be given in Chapters III and XI, where we show that, up to homotopy, "$R_\infty X$ is an Artin-Mazur-like R-completion of X" in two different senses.

It can be shown (see Ch. X, 4.9, 4.10 and 5.1) that any surjection $X \to Y \in \mathcal{S}$ induces a fibration $R_\infty X \to R_\infty Y$ and thus $R_\infty X$ is always fibrant. We will also often use the fact that

4.3 $R_\infty X$ is the inverse limit of a tower of fibrations $\{R_s X\}$. If for each $s \geq -1$, we put (3.4)

$$R_s X = \text{Tot}_s \underset{\sim}{R}X \qquad \in \mathcal{S}$$

then 2.2 and (Ch. X, 4.9 and 4.10) imply that $\{R_s X\}$ is a tower of fibrations such that

$$R_\infty X = \underset{\leftarrow}{\lim} R_s X .$$

Hence (Ch. IX, 3.1) there is, for every $i \geq 0$ and choice of base point $* \in X$, a short exact sequence

$$* \longrightarrow \lim_{\leftarrow}{}^{1} \pi_{i+1} R_s X \longrightarrow \pi_i R_\infty X \longrightarrow \lim_{\leftarrow} \pi_i R_s X \longrightarrow * \quad .$$

Another consequence is the existence of

4.4 The associated spectral sequence. For $X \in \mathcal{S}$, a choice of base point $* \in X$ makes $\underset{\sim}{R}X$ and hence the tower of fibrations $\{R_s X\}$ pointed. Thus (Ch. IX, 4.2) one can form the extended homotopy spectral sequence of this tower. It turns out (see Ch. X, 6.4) that in dimensions ≥ 1 this spectral sequence coincides with the homotopy spectral sequence $\{E_r^{s,t}(X;R)\}$ of X with coefficients in R of [Bousfield-Kan (HS)], which

(i) for $R = Z_p$ (the integers modulo a prime p) is "the" [Bousfield-Kan (HS), §1] unstable Adams spectral sequence, and

(ii) for $R = Q$ (the rationals) consists of the primitive elements in the rational cobar spectral sequence [Bousfield-Kan (PP), §15].

The convergence of this spectral sequence will be investigated in Ch.V, §3 and Ch.VI, §9.

We end with remarking that

4.5 The ring R can (and will) always be assumed to be "solid", i.e. the multiplication map $R \otimes_Z R \to R$ is an isomorphism. To be precise, let R be a commutative ring and let $cR \subset R$ be its core, i.e. the maximal solid subring of R, or equivalently [Bousfield-Kan (CR)] the subring given by

$$cR = \{x \in R \mid 1 \otimes x = x \otimes 1 \in R \otimes_{\mathbf{Z}} R\}.$$

Then we will prove in §9 the

9.1 Core lemma. *Let R be a commutative ring and let $cR \subset R$ be its core. Then the inclusion $cR \subset R$ induces, for every $X \in \mathcal{S}$, a homotopy equivalence*

$$(cR)_\infty X \simeq R_\infty X \qquad \in \mathcal{S} .$$

§5. R-complete, R-good and R-bad spaces

Depending on how much $R_\infty X$ resembles X one can consider three classes of spaces.

5.1 R-complete spaces, R-good spaces and R-bad spaces. A space $X \in \mathcal{S}$ is called

(i) R-complete if the map $\phi: X \to R_\infty X$ is a weak (homotopy) equivalence,

(ii) R-good if $\phi_*: \tilde{H}_*(X;R) \to \tilde{H}_*(R_\infty X;R)$ is an isomorphism, and

(iii) R-bad if it is not R-good.

Our main purpose here is to prove

5.2 Proposition. For a space $X \in \mathcal{S}$ the following conditions are equivalent:

(i) X is R-good,

(ii) $R_\infty X$ is R-complete,

(iii) $R_\infty X$ is R-good.

This implies that, roughly speaking, "a good space is very good and a bad space is very bad", i.e.

5.3 Corollary. For $X \in \mathcal{S}$, the sequence

$$R_\infty X \xrightarrow{\phi} R_\infty^2 X \longrightarrow \cdots \longrightarrow R_\infty^k X \xrightarrow{\phi} R_\infty^{k+1} X \longrightarrow \cdots$$

either "consists of only homotopy equivalences", or "contains no homotopy equivalence".

In Chapters V, VI and VII we give various examples of R-good spaces and we show there that "most" (but not all) spaces are R-good for $R \subset Q$ and $R = Z_p$. On the other hand, an infinite wedge of circles is Z_p-bad (Ch. IV, 5.4), while some finite wedge of circles and the projective plane P^2 are Z-bad (Ch. VII, §5).

Proof of proposition 5.2. This proposition is an easy consequence of the following lemmas, which are of some interest in their own right.

5.4 Lemma. *For every $X \in \mathcal{S}$, the map*

$$\phi_*: \tilde{H}_*(X;R) \longrightarrow \tilde{H}_*(R_\infty X;R)$$

has a natural left inverse, i.e. ϕ_* *is a monomorphism onto a natural direct summand.*

Proof. This lemma follows from 2.3 and the fact that the map

$$RX \xrightarrow{R\phi} RR_\infty X$$

has as a left inverse the composition

$$RR_\infty X \xrightarrow{R(\text{proj.})} RRX \xrightarrow{\psi} RX \quad .$$

5.5 Lemma. *A map $f: X \to Y \in \mathcal{S}$ induces an isomorphism*

$$f_*: \tilde{H}_*(X;R) \approx \tilde{H}_*(Y;R)$$

if and only if it induces a homotopy equivalence

$$R_\infty f\colon R_\infty X \simeq R_\infty Y \qquad\qquad \epsilon\ \mathcal{S}\ .$$

<u>Proof</u>. The "only if" part follows from §2 and Ch. X, 5.2 and the "if" part follows from 5.4.

An obvious consequence of 5.5 is that R_∞ <u>induces a functor from the (pointed) homotopy category of spaces to itself</u>.

The proof of proposition 5.2 can be completed using

<u>5.6 Triple lemma</u>. <u>There exist natural transformations</u>

$$\mathrm{Id} \xrightarrow{\phi} R_s \qquad \underline{\text{and}} \qquad R_s^2 \xrightarrow{\psi} R_s \qquad\qquad 0 \le s \le \infty$$

<u>such that</u>

<u>(i) for $s = 0$</u> (i.e. $R_s = R$) <u>ϕ and ψ are as in §2</u>,

<u>(ii) for $s = \infty$</u> (i.e. $R_s = R_\infty$) <u>ϕ is as in 4.2</u>,

<u>(iii)</u> $\{R_s,\phi,\psi\}$ <u>is a triple for all</u> $0 \le s \le \infty$, <u>and</u>

<u>(iv) these triples are compatible in the sense that the obvious diagrams</u>

$$\begin{array}{ccc} \mathrm{Id} & \xrightarrow{\phi} & R_s \\ {\scriptstyle =}\downarrow & & \downarrow \\ \mathrm{Id} & \xrightarrow{\phi} & R_{s'} \end{array} \qquad\qquad \begin{array}{ccc} R_s^2 & \xrightarrow{\psi} & R_s \\ \downarrow & & \downarrow \\ R_{s'}^2 & \xrightarrow{\psi} & R_{s'} \end{array}$$

<u>commute for all</u> $0 \le s' < s \le \infty$.

To prove this we need

<u>5.7 A characterization of triples</u>. Let $\mathcal{C}$ be a category,

let $T: \mathcal{C} \to \mathcal{C}$ be a functor and let

$$\phi: \mathrm{Id} \longrightarrow T \qquad \text{and} \qquad \psi: T^2 \longrightarrow T$$

be natural transformations such that $\{T,\phi,\psi\}$ is a triple. Then the pairing c which assigns to every pair of maps

$$f: X \longrightarrow TY, \qquad g: W \longrightarrow TX \qquad\qquad \varepsilon\ \mathcal{C}$$

the composition

$$c(f,g): W \xrightarrow{g} TX \xrightarrow{Tf} T^2Y \xrightarrow{\psi Y} TY \qquad\qquad \varepsilon\ \mathcal{C}$$

clearly has the properties

(i) c is natural (in an obvious sense)

(ii) c is associative

(iii) for every map $f: X \to TY$ $\varepsilon\ \mathcal{C}$

$$c(f,\phi X) \;=\; f \;=\; c(\phi Y,f) \ .$$

Conversely, given T,ϕ and a pairing c with these three properties, one can, for every object $Y \varepsilon \mathcal{C}$, define a map

$$\psi Y \;=\; c(\mathrm{id},\mathrm{id}): T^2Y \longrightarrow TY \qquad\qquad \varepsilon\ \mathcal{C}$$

and a straightforward calculation then yields that *the function ψ so defined is, in fact, a natural transformation $T^2 \to T$, and that $\{T,\phi,\psi\}$ is a triple*.

Proof of triple lemma 5.6. For $Y \varepsilon \mathcal{A}$, let

$$(\underset{\sim}{R}Y)^{k-1} \;=\; R^kY \xrightarrow{t_{k-i}} R^kY \;=\; (\underset{\sim}{R}Y)^{k-1}$$

be the twist map which "interchanges the (k-i)-th and (k-i-1)-th copies of R (counted from Y)", i.e. (see 2.4)

$$t_{k-i} = d^i s^i + d^{i+1} s^i - \mathrm{id}$$

and let

$$(\underset{\sim}{R}Y)^{2n-1} = R^{2n}Y \xrightarrow{w_n} R^n Y = (\underset{\sim}{R}Y)^{n-1}$$

be the map which "combines the i-th and (n+i)-th copies of R^{2n}", i.e. w_n is the composition

$$R^{2n}Y \xrightarrow{t_{n+1}} \cdots \xrightarrow{t_{2n-1}} R^{2n}Y \xrightarrow{s^0} R^{2n-1}Y \xrightarrow{Rw_{n-1}} R^n Y$$

where $w_1 = s^0$. For $W, X, Y \in \mathscr{A}$ one can then form the map of cosimplicial spaces

$$c\colon \hom(X,\underset{\sim}{R}Y) \times \hom(W,\underset{\sim}{R}X) \longrightarrow \hom(W,\underset{\sim}{R}Y)$$

which assigns to a pair of q-simplices

$$u\colon \Delta[q] \times X \longrightarrow R^n Y \qquad \in \hom(X,\underset{\sim}{R}Y)^{n-1}_q$$

$$v\colon \Delta[q] \times W \longrightarrow R^n X \qquad \in \hom(W,\underset{\sim}{R}X)^{n-1}_q$$

the composition

$$c(u,v)\colon \Delta[q] \times W \longrightarrow \Delta[q] \times \Delta[q] \times W \longrightarrow \Delta[q] \times R^n X$$
$$\longrightarrow R^n(\Delta[q] \times X) \longrightarrow R^{2n}Y \xrightarrow{w_n} R^n Y \qquad \in \mathscr{A}$$

where the unnamed maps are the obvious ones. Of course one has to verify that c is indeed a cosimplicial map, but that is straightforward (although not short). Moreover it is not hard to see that this map c induces a pairing of function spaces

$$c: \hom(X, R_s Y) \times \hom(W, R_s X) \longrightarrow \hom(W, R_s Y) \qquad \in \mathcal{S}$$

which in dimension 0 has the three properties of 5.7. Hence the function ψ given by

$$\psi Y = c(\mathrm{id}, \mathrm{id}): R_s^2 Y \longrightarrow R_s Y$$

is a natural transformation such that $\{R_s, \phi, \psi\}$ is a triple. The rest of the lemma now is easy.

§6. Low dimensional behavior

In spite of the fact that, in general, the k-skeleton $(R_\infty X)^{[k]}$ is not contained in any of the spaces $R_\infty(X^{[n]})$, we will show that "*the homotopy type of* $R_\infty X$ *in dimensions* $< k$" depends only on (part of) "*the homotopy type of* X *in dimensions* $\leq k$". More precisely

6.1 Connectivity lemma. *Let* R *be a solid ring (4.5), let* $k \geq 0$ *and let* $X \in \mathcal{S}$ *be such that* $\tilde{H}_i(X;R) = 0$ *for* $i \leq k$. *Then*

(i) the fibres of the maps $R_s X \to R_{s-1} X$ *are k-connected for all* $s < \infty$, *and hence (Ch. IX, 3.1)*

(ii) the space $R_\infty X$ *is k-connected.*

6.2 Relative connectivity lemma. *Let* R *be a solid ring (4.5), let* $k \geq 0$ *and let* $f\colon X \to Y \in \mathcal{S}$ *be such that the induced map* $\tilde{H}_i(X;R) \to \tilde{H}_i(Y;R)$ *is an isomorphism for* $i \leq k$ *and is onto for* $i = k+1$. *Then, for every choice of base point,*

(i) the induced maps $\pi_i R_s X \to \pi_i R_s Y$ $(s < \infty)$ *are isomorphisms for* $i \leq k$ *and onto for* $i = k+1$, *and hence* (Ch. IX, 3.1)

(ii) The induced map $\pi_i R_\infty X \to \pi_i R_\infty Y$ *is an isomorphism for* $i < k$ *and is onto for* $i = k$.

A somewhat stronger version of 6.2 (ii) will be obtained in Ch. IV, 5.1.

6.3 Corollary. *Let* $X \in \mathcal{S}$ *be fibrant (i.e.* $X \to *$ *is a fibration) and let*

$$X \longrightarrow \cdots \longrightarrow X^{(k)} \longrightarrow X^{(k-1)} \longrightarrow \cdots$$

<u>denote its Postnikov tower [May, p. 33]. Then the induced map</u>

$$R_\infty X \longrightarrow \underset{\leftarrow}{\lim}\, R_\infty X^{(k)}$$

<u>is a homotopy equivalence</u>.

<u>6.4 Remark on the solidity of R</u>. The first part of lemmas 6.1 and 6.2 is only true for <u>solid</u> rings, but parts (ii) hold, of course (4.5), without this restriction.

<u>Proof of 6.1</u>. Choose a base point $* \in X$. In view of Ch. X, 6.2, it then suffices to show that each $N\underset{\sim}{R}X^s$ is (k+s)-connected, where (see 2.2)

$$N\underset{\sim}{R}X^s = \underset{\sim}{R}X^s \cap \ker s^0 \cap \cdots \cap \ker s^{s-1} \qquad \subset \underset{\sim}{R}X^s.$$

To do this consider the functors

$$T^s\colon (\underline{R\text{-modules}}) \longrightarrow (\underline{R\text{-modules}})$$

given by (see 2.2)

(i) $T^0M = M$

(ii) $T^1M = \ker \psi\colon RM \to M$, where ψ denotes the homomorphism given by $1m \to m$ for all $m \in M$, and

(iii) $T^sM = T^{s-1}T^1M \oplus T_2^{s-1}(M,T^1M)$, where T_2^{s-1} denotes the 2-fold <u>cross effect</u> of T^{s-1}, i.e.

$$T_2^{s-1}(M',M'') = \ker (T^{s-1}(M'\oplus M'') \longrightarrow T^{s-1}M' \oplus T^{s-1}M'').$$

Since there are natural isomorphisms

$$T^sRX \approx \ker (T^{s-1}R^2X \xrightarrow{T^{s-1}\psi} T^{s-1}RX) \qquad s \geq 1$$

it is easy to see that there are natural isomorphisms

$$T^s RX \approx N\underset{\sim}{R}X^s$$

The desired result now follows readily, by induction on s, from the fact that T^1RX is $(k+1)$-connected (because R is solid) and the following

6.5 Lemma [Curtis (L), §5]. Let $k \geq 0$ and let

$$T: (R\text{-modules}) \longrightarrow (R\text{-modules})$$

be a functor which commutes with direct limits and is such that $T0 = 0$ and that, for every connected simplicial free R-module B, the space TB is k-connected. Then, for every m-connected simplicial free R-module C $(m \geq 0)$, the space TC is $(m+k)$-connected.

Proof of 6.2. To prove 6.2 we will use the disjoint union lemma 7.1. This is permissible, as the proof of 7.1 involves 6.1, but not 6.2. We clearly may assume that f is onto, and, in view of 7.1, that X and Y are connected.

Using the notation of the proof of 6.1 it thus suffices to show first that the induced map $\pi_i T^s RX \to \pi_i T^s RY$ is an isomorphism for $i \leq k+s$ and is onto for $i = k+s+1$. To do this we recall from [Kan-Whitehead, §16] and [Curtis (L), §5] the existence of a magic exact sequence of simplicial R-modules

$$\cdots \longrightarrow T^s_j(K,\cdots,K) \oplus T^s_{j+1}(K,\cdots,K,RX) \longrightarrow \cdots$$

$$\cdots \longrightarrow T^s K \oplus T^s_2(K,RX) \longrightarrow T^s RX \longrightarrow T^s RY \longrightarrow *$$

where $K = \ker (Rf\colon RX \to RY)$ and T^s_j denotes the j-fold cross effect of the functor T^s. As K is k-connected (because f is onto), lemma 6.5 (together with the fact that R is solid) readily implies that in the above magic exact sequence all spaces, except possibly T^sRX and T^sRY, are $(k+s)$-connected. This proves the desired result.

§7. Disjoint unions, finite products and multiplicative structures

We now state and prove the not very surprising results that, up to homotopy, the R-completion functor commutes with *disjoint unions* and *finite products*, and preserves *multiplicative structures*.

7.1 Disjoint union lemma. *Let* $X \in \mathcal{S}$ *and for each* $a \in \pi_0 X$ *let* $X_a \subset X$ *denote the corresponding component. Then the inclusion of the disjoint union*

$$\coprod_a R_\infty X_a \longrightarrow R_\infty X \qquad \in \mathcal{S}$$

is a homotopy equivalence.

It should be noted that the *pointed* version of this lemma is *false*, even in the *finite* case, as *some finite wedge of circles is not Z-good* (Ch.VII, 5.3), while *the circle itself is Z-complete*.

7.2 Finite product lemma. *For* $X, Y \in \mathcal{S}$, *the projections of* $X \times Y$ *onto* X *and* Y *induce a homotopy equivalence*

$$R_\infty(X \times Y) \simeq R_\infty X \times R_\infty Y \qquad \in \mathcal{S} .$$

Moreover this map has a natural left inverse

$$a\colon R_\infty X \times R_\infty Y \longrightarrow R_\infty(X \times Y) \qquad \in \mathcal{S}$$

which is also associative, commutative and compatible with the triple structure of R_∞ (5.6).

7.3 Corollary. A multiplication

$$m\colon X \times X \longrightarrow X \qquad \varepsilon\ \mathcal{S}$$

induces a multiplication

$$m'\colon R_\infty X \times R_\infty X \xrightarrow{a} R_\infty(X \times X) \xrightarrow{R_\infty m} R_\infty X \qquad \varepsilon\ \mathcal{S} \ .$$

Moreover, if m is associative, commutative or has a left (or right) unit, then so does m'.

7.4 Remark. If m has a unit and an inverse, then m need not have an inverse, as the following triangle commutes, in general, only up to homotopy

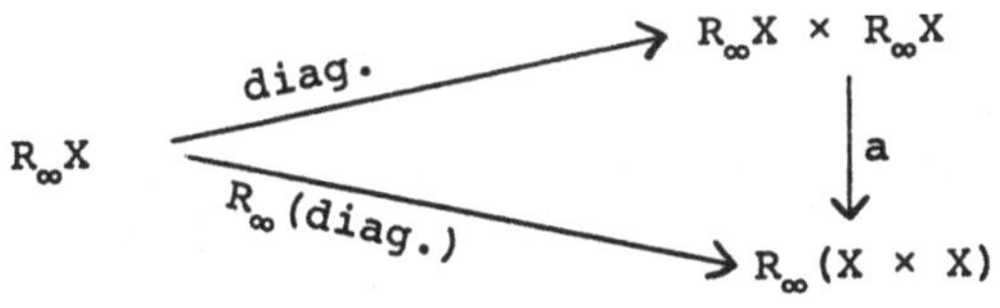

7.5 Corollary. Let X be an H-space, i.e. X has a base point $*$ and a pointed multiplication map $m\colon X \times X \to X \ \varepsilon\ \mathcal{S}$, such that, in the pointed homotopy class $[X,X]$ (Ch.IX, §3)

$$m(\mathrm{id},*) = \mathrm{id} = m(*,\mathrm{id}).$$

Then $R_\infty X$ is also an H-space. Moreover, if X is, for instance, homotopy associative or homotopy commutative, then so is $R_\infty X$.

The lemmas 7.1 and 7.2 will be proved using the theory of

7.6 Acyclic models [Barr-Beck]. Given a category $\mathcal{C}$, a functor $T: \mathcal{C} \to \mathcal{C}$, a natural transformation $e: \mathrm{Id} \to T$ and a cochain functor K, i.e. functors

$$K^n: \mathcal{C} \longrightarrow \text{(abelian groups)} \qquad n \geq -1$$

and natural transformations $d: K^n \to K^{n+1}$ such that $dd = 0$, one says that

(i) K is T-acyclic if there is a natural contracting homotopy for the composite cochain functor KT, and that

(ii) K is T-representable if there are natural transformations

$$t^n: K^nT \longrightarrow K^n \qquad n \geq 0$$

such that

$$K^n \xrightarrow{K^n e} K^nT \xrightarrow{t^n} K^n$$

is the identity. Then [Barr-Beck] prove

7.7 Lemma. Let K be a cochain functor on $\mathcal{C}$ which is T-acyclic and let L be a cochain functor which is T-representable. Then any natural transformation $f^{-1}: K^{-1} \to L^{-1}$ can be extended to a natural cochain map $f: K \to L$. Moreover, if $f,g: K \to L$ are natural cochain maps such that $f^{-1} = g^{-1}$, then there exists a natural cochain homotopy $f \simeq g$.

In our proofs of lemmas 7.1 and 7.2 we will use

7.8 A slight generalization. The acyclic model lemma 7.7 also works for cochain functors K, L which are non-abelian in dimension -1, i.e.

$$K^{-1}, L^{-1}: \mathcal{C} \longrightarrow \text{(groups)}.$$

Proof of 7.1. For each $a \in \pi_0 X$ choose a base point $*_a \in X_a$. In view of 2.2, each such choice makes $\underset{\sim}{R}X$ and $\underset{\sim}{R}\pi_0 X$ group-like (Ch.X, 4.8) and thus gives rise to a group-like cosimplicial space

$$\underset{\sim}{R}(X;a) = \ker(\underset{\sim}{R}X \xrightarrow{\text{proj.}} \underset{\sim}{R}\pi_0 X) \subset \underset{\sim}{R}X .$$

One readily verifies that the inclusion of the disjoint union

$$\coprod_a \operatorname{Tot}_s \underset{\sim}{R}(X;a) \longrightarrow \operatorname{Tot}_s \underset{\sim}{R}X = R_s X$$

is an isomorphism for all $1 \leq s \leq \infty$, and it thus remains to show that, for each $a \in \pi_0 X$, the inclusion

$$R_\infty X_a = \operatorname{Tot} \underset{\sim}{R}X_a \longrightarrow \operatorname{Tot} \underset{\sim}{R}(X;a)$$

is a homotopy equivalence. In view of 6.1, 4.4 and Ch.X, 7.1 and 7.7, one thus has to prove that (in the notation used there) the cochain maps

$$(\pi_t \underset{\sim}{R}X_a, d) \longrightarrow (\pi_t \underset{\sim}{R}(X;a), d) \qquad t \geq 1$$

are cochain homotopy equivalences. This we will do using 7.7 and 7.8.

Let $\mathcal{S}_*$ denote the category of spaces with base point and let $\mathcal{C} \subset \mathcal{S}_*$ be the subcategory consisting of the maps for which π_0 is 1-1 (but not necessarily onto). Let $T = R$ and let $e = \phi$. Then a simple calculation (or [Bousfield-Kan (HS), 4.4]) yields that both cochain functors are T-acyclic. Moreover the fact that the functor R admits a triple structure (§2) implies readily that the second cochain functor is T-representable, while the T-representability of the first is an easy consequence of the fact that $\pi_t RX_a = \tilde{H}_t(X_a;R)$ is a natural direct summand of $\pi_t RX \approx \tilde{H}_t(X;R)$. (This is not true on all of $\mathcal{A}_*$).

Application of 7.7 and 7.8 (several times) now yields the desired result.

Proof of 7.2. In view of 7.1 we can assume that X and Y are connected and hence (6.1, 4.4 and Ch.X, 7.1 and 7.7) we have to show that, for every choice of base points in X and Y, the cochain maps

$$(\pi_t \underset{\sim}{R}(X \times Y),d) \longrightarrow (\pi_t(\underset{\sim}{R}X \times \underset{\sim}{R}Y),d) \qquad t \geq 1$$

are cochain homotopy equivalences. This we again, as in the proof of 7.1, do using 7.7 and 7.8.

Let $\mathcal{C} = \mathcal{A}_* \times \mathcal{A}_*$, where $\mathcal{A}_*$ is as in the proof of 7.1, let $T = R \times R$ and let $e = \phi \times \phi$. Then, as in the proof of 7.1, one readily verifies that both cochain functors are T-acyclic and that the second one is also T-representable, while the T-representability of the first one follows from the fact that the map

$$R(X \times Y) \xrightarrow{R(\phi \times \phi)} R(RX \times RY)$$

has a natural left inverse, namely the homomorphism given by the formula

$$(\sum_i r_i x_i, \sum_j r'_j y_j) \longrightarrow \sum_{i,j} r_i r'_j (x_i, y_j)$$

To prove the rest of the lemma observe that this formula actually defines a map

$$\alpha: RX \times RY \longrightarrow R(X \times Y).$$

This can be used to construct a cosimplicial map

$$\underset{\sim}{R}X \times \underset{\sim}{R}Y \longrightarrow \underset{\sim}{R}(X \times Y) \quad \in \quad c\mathcal{S}$$

which in codimension n is the composition

$$R^{n+1}X \times R^{n+1}Y \xrightarrow{\alpha} R(R^nX \times R^nY) \xrightarrow{R\alpha} \cdots \xrightarrow{R^n\alpha} R^{n+1}(X \times Y) \ .$$

Taking total spaces one then gets the desired map

$$a\colon R_\infty X \times R_\infty Y \longrightarrow R_\infty(X \times Y) \qquad \in \mathcal{S} \ .$$

§8. The fibre-wise R-completion

The notion of R-completion will now be generalized to a notion of fibre-wise R-completion, i.e. we will, for a fibration $X \to B \in \mathcal{S}$, construct in a functorial manner a fibration $\dot{R}_\infty X \to B \in \mathcal{S}$, of which the fibres are the R-completions of the fibres of the map $X \to B$.

8.1 Construction of the fibre-wise R-completion. We start with generalizing the functor $R: \mathcal{S} \to \mathcal{S}$. For a map $f: X \to B \in \mathcal{S}$ (which need not be a fibration) let

$$\dot{R}X \subset RX$$

denote the subspace consisting of the simplices

$$r_1x_1 + \cdots + r_kx_k \qquad x_i \in X,\ r_i \in R,\ \Sigma\, r_i = 1$$

for which all x_i lie over the same simplex of B, i.e.

$$fx_1 = \cdots = fx_k .$$

There is an obvious map $\dot{R}f: \dot{R}X \to B \in \mathcal{S}$ and hence one can repeat this construction and obtain subspaces

$$\dot{R}^n X \subset R^n X \qquad n \geq 0$$

which together yield a cosimplicial subspace

$$\dot{\underset{\sim}{R}}X \subset \underset{\sim}{R}X .$$

Now we put

$$\dot{R}_s X = \operatorname{Tot}_s \dot{\underset{\sim}{R}} X \qquad s \le \infty.$$

The desired *fibre-wise R-completion* then is

$$\dot{R}_\infty X = \underset{\leftarrow}{\lim}\, \dot{R}_s X \quad .$$

There are obvious commutative diagrams

$$\begin{array}{ccccc} X & \longrightarrow & \dot{R}_s X & \longrightarrow & R_s X \\ \downarrow & & \downarrow & & \downarrow \\ B & \xrightarrow{=} & B & \longrightarrow & R_s B \end{array} \qquad s \le \infty$$

in which the square on the right is, in general, *not* a pull back. It is also not hard to see that

(i) If $B = *$, then $\dot{R}_s X = R_s X \quad (s \le \infty)$

(ii) The construction is natural, i.e. a commutative diagram

$$\begin{array}{ccc} X & \longrightarrow & Y \\ \downarrow & & \downarrow \\ B & \longrightarrow & C \end{array}$$

gives rise to commutative diagrams

$$\begin{array}{ccc} \dot{R}_s X & \longrightarrow & \dot{R}_s Y \\ \downarrow & & \downarrow \\ B & \longrightarrow & C \end{array} \qquad s \le \infty$$

(iii) If the first diagram in (ii) is a pull back, then so is the second for all $s \le \infty$.

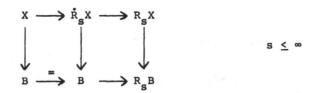

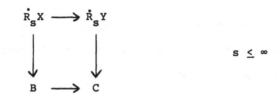

More difficult to prove is

<u>8.2 Proposition. If $f: X \to B \in \mathscr{S}$ is a fibration, then so are the induced maps $\dot{R}_s X \to B \in \mathscr{S}$ for all $s \leq \infty$.</u>

And combining this with 8.1 (i) and (iii) one gets

<u>8.3 Corollary. Let $X \to B \in \mathscr{S}$ be a pointed fibration with fibre F. Then $R_s F$ is the fibre of the map $\dot{R}_s X \to B$ $(s \leq \infty)$.</u>

<u>Proof of 8.2.</u> In view of Ch.X, 4.6, it suffices to show that (in the notation used there) the maps

$$\dot{R}X \xrightarrow{\dot{R}f} B \qquad \text{and} \qquad \dot{R}^{n+2}X = (\underset{\sim}{\dot{R}X})^{n+1} \xrightarrow{s} M^n \underset{\sim}{\dot{R}X}$$

are fibrations, and this can be done as follows.

As $f: X \to B$ is a fibration one can, for every pair of integers (i,n) with $0 \leq i \leq n$ and every n-simplex $b \in B$, choose in X <u>functions</u>

$$s_{i,b}: f^{-1}(d_i b) \longrightarrow f^{-1}(b)$$

<u>such that</u> $d_i s_{i,b} = \text{id}$. These functions induce in $\dot{R}X$ similar functions

$$s_{i,b}: (\dot{R}f)^{-1}(d_i b) \longrightarrow (\dot{R}f)^{-1}(b)$$

and the proof that $\dot{R}f: \dot{R}X \to B$ is a fibration now is essentially the same as the proof that a simplicial group is fibrant (see, for instance, [May, p. 67]), except that one uses the functions $s_{i,b}$ instead of the degeneracies s_i.

The proof that the map $s: \dot{R}^{n+2}X \to M^n \underset{\sim}{\dot{R}X}$ is a fibration is similar and uses the facts that

(i) the map $s\colon \dot{R}^{n+2}X \to M^n\dot{\underset{\sim}{R}}X$ is onto,

(ii) the functions $s_{i,b}$ in X induce similar functions in $\dot{R}^{n+2}X$ and $M^n\dot{\underset{\sim}{R}}X$, and

(iii) the map $s\colon \dot{R}^{n+2}X \to M^n\dot{\underset{\sim}{R}}X$ is compatible with the functions $s_{i,b}$.

The last two of these statements are easily verified, while the proof of the first one is as in Ch.X, 4.9.

§9. The role of the ring R

We end this chapter with an investigation of the role of the ring R and show that for a large class of rings the homotopy type of $R_\infty X$ is completely determined by the homotopy types of $R_\infty X$ for $R = Z_p$ (the integers modulo a prime p) and $R \subset Q$ (subring of the rationals).

We start with observing (see 4.5) that one only has to consider solid rings, i.e. rings for which the multiplication map $R \otimes_Z R \to R$ is an isomorphism. More precisely, if R is a commutative ring and $cR \subset R$ its core, i.e. its maximal solid subring or equivalently [Bousfield-Kan (CR)] the subring

$$cR = \{x \in R \mid 1\otimes x = x\otimes 1 \in R \otimes_Z R\}$$

then we have the

9.1 Core lemma. Let R be a commutative ring and let $cR \subset R$ be its core. Then the inclusion $cR \to R$ induces, for every $X \in \mathcal{S}$, a homotopy equivalence

$$(cR)_\infty X \simeq R_\infty X \qquad \in \mathcal{S}.$$

In fact this reduction also holds for

9.2 Certain non-commutative rings. The definition of R-completion (§2 and §4) clearly also makes sense for non-commutative rings. All one has to do is, replace everywhere R-module by left R-module. It is, however, questionable whether this gives anything new, as an analysis of the proof of 9.1 yields:

Let R be a commutative ring and R' a not necessarily commutative ring for which there exists an abelian group homomorphism $R' \to R$ which sends 1 into 1. Then any ring homomorphism $R \to R'$ induces, for every $X \in \mathcal{J}$, a homotopy equivalence

$$R_\infty X \simeq R'_\infty X \qquad \in \mathcal{J} .$$

For example, this is the case if $R' = R[\pi]$, the group ring of a group π over the commutative ring R.

The next step is

9.3 Determination of all solid rings. This was done in [Bousfield-Kan (CR)], and we recall from there that the only solid rings are

(i) the cyclic rings Z_n for $n \geq 2$,

(ii) the subrings of the rationals, i.e. the rings $Z[J^{-1}]$ for any set J of primes, where $Z[J^{-1}]$ consists of those rationals whose denominators involve only primes in J,

(iii) the product rings $Z[J^{-1}] \times Z_n$, where each prime factor of n is in J, and

(iv) all direct limits (over directed systems) of the above three types of rings.

Finally we state two propositions which imply that the homotopy types of $R_\infty X$ for $R = Z_p$ (p prime) and $R \subset Q$ completely determine the homotopy type of $R_\infty X$ for any solid ring of type (i), (ii) or (iii) above.

9.4 Proposition. Let $R = Z_p$ (p prime) and let $R' = Z_{p^n}$. Then the projection $Z_{p^n} \to Z_p$ induces, for every $X \in \mathcal{J}$, *a homotopy equivalence*

$$R'_\infty X \simeq R_\infty X \qquad \in \mathcal{S} .$$

9.5 Proposition. *Let either*

(i) $R = Z_m$ *and* $R' = Z_n$, *where* m *and* n *are integers such that* $(m,n) = 1$, *or*

(ii) $R = Z[J^{-1}]$ *and* $R' = Z_n$, *where* n *is an integer of which all the prime factors are in* J *(see 9.3 (iii))*.

Then the projections of $R \times R'$ *onto* R *and* R' *induce, for every* $X \in \mathcal{S}$, *a homotopy equivalence*

$$(R \times R')_\infty X \simeq R_\infty X \times R'_\infty X \qquad \in \mathcal{S} .$$

Proofs. In view of 7.1 it suffices to prove 9.1, 9.4 and 9.5 for *connected* X.

To prove 9.1 one combines 6.1 with 4.4, Ch.X, 7.4 and the fact that [Bousfield-Kan (CR)] the inclusion $cR \to R$ induces, for every choice of base point, an isomorphism

$$E_2(X;cR) \approx E_2(X;R)$$

The proof of 9.5 is similar and uses the fact that [Bousfield-Kan (HS), §8] the projections of $R \times R'$ onto R and R' induce, for every choice of base point, an isomorphism

$$E_2(X;R \times R') \approx E_2(X;R) \oplus E_2(X;R')$$

And finally to prove 9.4 one observes that there are natural isomorphisms

$$\operatorname{Tot} \operatorname{Tot}^{(1)} \underset{\sim}{R}\underset{\sim}{R}{}' X \approx \operatorname{Tot} \operatorname{Tot}^{(2)} \underset{\sim}{R}\underset{\sim}{R}{}' X$$

where $\mathrm{Tot}^{(1)}$ and $\mathrm{Tot}^{(2)}$ denote the first and second cosimplicial total space of the double cosimplicial space $\underset{\sim}{R}\underset{\sim}{R}'X$. Moreover it follows readily from [Bousfield-Kan (HS), 10.6] and Ch.VI, 5.4 that the natural maps

$$R'_\infty X = \mathrm{Tot}\, \underset{\sim}{R}'X \longrightarrow \mathrm{Tot}\, \mathrm{Tot}^{(1)} \underset{\sim}{R}\underset{\sim}{R}'X \qquad \in \mathcal{J}$$

$$R_\infty X = \mathrm{Tot}\, \underset{\sim}{R}X \longrightarrow \mathrm{Tot}\, \mathrm{Tot}^{(2)} \underset{\sim}{R}\underset{\sim}{R}'X \qquad \in \mathcal{J}$$

are homotopy equivalences. The rest of the proof now is easy.

Chapter II. Fibre lemmas

§1. Introduction

For a general fibration of connected spaces $F \to E \to B$, the map $R_\infty E \to R_\infty B$ is always a fibration (Ch.I, 4.2), but $R_\infty F$ need not have the same homotopy type as the fibre of $R_\infty E \to R_\infty B$. For example, if $R = Q$, then

$$S^2 \longrightarrow P^2 \longrightarrow K(Z_2,1)$$

is, up to homotopy, a fibration, but $R_\infty S^2 \to R_\infty P^2 \to R_\infty K(Z_2,1)$ is not, because (Ch.I, 5.5) $R_\infty P^2$ and $R_\infty K(Z_2,1)$ are contractible, while $R_\infty S^2$ is not.

However, we will prove in this chapter a mod-R fibre lemma (5.1) which, roughly speaking, states that the R-completion preserves, up to homotopy, fibrations of connected spaces $F \to E \to B$, for which "$\pi_1 B$ acts nilpotently on each $\tilde{H}_i(F; R)$". This is a useful result, for instance, when one wants to compute $\pi_* R_\infty X$ in terms of $\pi_* X$, using Postnikov methods.

We obtain the mod-R fibre lemma in several steps as follows:

§2 and §3 In §2 we state a special case, the principal fibration lemma, and obtain several consequences thereof. A rather long and technical proof of this principal fibration lemma is the content of §3.

§4 introduces the notion of nilpotent fibration and we prove here, with the use of the principal fibration lemma, a more general nilpotent fibration lemma.

§5 and §6 In §5 we finally state the mod-R fibre lemma and discuss various special cases, while §6 contains a proof which uses the nilpotent fibration lemma of §4 and the fibre-wise R-completion of Ch.I, §8. A different proof will be given in Chapter III, §7.

Notational warnings. Throughout most of this chapter we will work in the category $\mathcal{S}_{*c}$ of pointed connected spaces. This is no real restriction as the R-completion commutes, up to homotopy, with disjoint unions (Ch.I, 7.1).

Of course (Ch.I, 4.5) we assume throughout that the ring R is solid.

§2. The principal fibration lemma

We start with a special case of the mod-R fibre lemma, the principal fibration lemma, which states that, *up to homotopy, the R-completion preserves principal fibrations with connected fibres*. We also list some corollaries and show that *the spaces* R_sX *$(s < \infty)$ are R-complete for all* $X \in \mathcal{S}_{*c}$ (the category of *pointed connected* spaces).

We first recall the definition of [May, p.70]:

2.1 Principal fibrations. Let $E \in \mathcal{S}_*$ (the category of *pointed* spaces), let $F \in \mathcal{S}_*$ be a *simplicial group* and let

$$a\colon F \times E \longrightarrow E \qquad\qquad \in \mathcal{S}_*$$

be a *principal action* (see 3.1). Then [May, p.70] the projection

$$p\colon E \longrightarrow B \;=\; E/\text{action} \qquad\qquad \in \mathcal{S}_*$$

is a fibration, which is called a *principal fibration*, with fibre F, as one can identify the fibre $p^{-1}*$ with F under the correspondence

$$a(f,*) \longleftrightarrow f \qquad\qquad f \in F\ .$$

More generally, we will call a map $f \in \mathcal{S}_*$ a *principal fibration, up to homotopy*, if f is equivalent in the pointed homotopy category (Ch.VIII, 4.6) to some principal fibration. By [May, Ch.IV and Ch.VI] this is the same as requiring that f be equivalent in the pointed homotopy category to an induced fibration of a path fibration over a connected space.

Now we state the

2.2 Principal fibration lemma. Let $p: E \to B \in \mathcal{J}_{*C}$ be a principal fibration with connected fibre F. Then $R_\infty p: R_\infty E \to R_\infty B \in \mathcal{J}_{*C}$ is a fibration which is, up to homotopy, also principal and the inclusion

$$R_\infty F = R_\infty(p^{-1}*) \longrightarrow (R_\infty p)^{-1}* \qquad \in \mathcal{J}_{*C}$$

is a homotopy equivalence.

This will be proved in §3.

2.3 Corollary. Let

$$E_n \longrightarrow E_{n-1} \longrightarrow \cdots \longrightarrow E_0 \qquad \in \mathcal{J}_{*C}$$

be a finite sequence of principal fibrations with connected fibres and let $p: E_n \to E_0 \in \mathcal{J}_{*C}$ be the composite fibration. Then $R_\infty p: R_\infty E_n \to R_\infty E_0 \in \mathcal{J}_{*C}$ is also a fibration and the inclusion $R_\infty(p^{-1}*) \to (R_\infty p)^{-1}* \in \mathcal{J}_{*C}$ is a homotopy equivalence.

Combining this with Ch.I, 6.3 one gets

2.4 Corollary. Let

$$\cdots \longrightarrow E_n \longrightarrow \cdots \longrightarrow E_0 \qquad \in \mathcal{J}_{*C}$$

be a tower of principal fibrations with fibres that are connected and that "get higher and higher connected" and let

$$p: E_\infty = \lim_{\leftarrow} E_n \longrightarrow E_0 \qquad \in \mathcal{J}_{*C}$$

be the composite fibration. Then $R_\infty p\colon R_\infty E_\infty \to R_\infty E_0 \in \mathcal{S}_{*c}$ *is also a fibration and the inclusion* $R_\infty(p^{-1}*) \to (R_\infty p)^{-1}* \in \mathcal{S}_{*c}$ *is a homotopy equivalence.*

Another consequence is

2.5 Proposition. *Let* $X \in \mathcal{S}_{*c}$. *Then the spaces* $R_s X$ $(s < \infty)$ *are R-complete*, i.e. the maps $\phi\colon R_s X \to R_\infty R_s X$ are homotopy equivalences.

This is an immediate consequence of 2.2 and the following two lemmas:

2.6 Lemma. *Let* $X \in \mathcal{S}_{*c}$. *Then the maps* $R_s X \to R_{s-1} X$ $(s < \infty)$ *are principal fibrations whose fibres are connected simplicial R-modules.*

Proof. This follows readily from Ch.I, 6.1 (R is solid) and the fact that (Ch.I, 2.2 and Ch.X, 4.10) the choice of base point makes $\tilde{R}X$ "R-module-like". By Ch.X, 6.2, the map $R_s X \to R_{s-1} X$ actually has $\hom_*(S^n, N\tilde{R}X^s)$ as fibre.

2.7 Lemma. *Every simplicial R-module B is R-complete.*

Proof. It suffices to show that one has in the homotopy spectral sequence (Ch.I, 4.4)

$$E_2^{0,t}(B;\, R) \approx \pi_t B$$

$$E_2^{s,t}(B;\, R) = * \qquad \text{for } s > 0.$$

This collapsing is not hard to prove, using the obvious homomorphism $RB \to B$ [Bousfield-Kan (HS), §10].

Using 2.6 one can also prove

2.8 Proposition. *Let* $f: X \to Y \in \mathcal{S}_{*C}$ *induce an isomorphism* $H_*(X; R) \approx H_*(Y; R)$. *Then* f *induces, for every* $W \in \mathcal{S}_{*C}$, *an isomorphism of pointed homotopy classes of maps*

$$[Y, R_\infty W] \approx [X, R_\infty W].$$

§3. Proof of the principal fibration lemma

To prove the principal fibration lemma 2.2 we use

3.1 A magic exact sequence. Let F be a group, E a pointed set and $a: F \times E \to E$ an action (i.e. $a(1,e) = e$ for all $e \in E$ and $a(f_1,a(f_2,e)) = a(f_1f_2,e)$ for all $f_1,f_2 \in F$ and $e \in E$). Then one can form an augmented space X (with base point) by putting

$$X_{-1} = E/\text{action}$$

$$X_0 = E$$

$$X_k = F \times \cdots \times F \times E \qquad k > 0$$

and defining faces and degeneracies by the formulas

$$d_0(f_1,\cdots,f_k,e) = (f_2f_1^{-1},\cdots,f_kf_1^{-1},a(f_1,e))$$

$$d_i(f_1,\cdots,f_k,e) = (f_1,\cdots,\hat{f}_i,\cdots,f_k,e) \qquad 0 < i \leq k$$

$$s_i(f_1,\cdots,f_k,e) = (f_1,\cdots,f_i,*,f_{i+1},\cdots,f_k,e) \qquad 0 \leq i \leq k.$$

A covariant functor

$$U: (\text{pointed sets}) \longrightarrow (\text{pointed sets})$$

such that $U* = *$ then can be applied to X dimensionwise and the resulting augmented space UX gives rise to a magic sequence

$$\cdots \longrightarrow U_kX \xrightarrow{d_0} U_{k-1}X \longrightarrow \cdots \longrightarrow U_{-1}X \longrightarrow *$$

where

$$U_k X \;=\; UX_k \cap \ker d_1 \cap \cdots \cap \ker d_k$$

and <u>short sequences</u>

$$* \longrightarrow U'_k X \longrightarrow U_k X \longrightarrow U'_{k-1} X \longrightarrow *$$

$$* \longrightarrow U^{k+1}(F,\cdots,F,E) \longrightarrow U_k X \longrightarrow U^k(F,\cdots,F) \longrightarrow *$$

where

$$U'_k X \;=\; UX_k \cap \ker d_0 \cap \cdots \cap \ker d_k \qquad \subset \qquad U_k X$$

and U^k denotes the k-fold <u>multiplicative</u> cross effect, i.e.

$$U^k(Y_1,\cdots,Y_k) = \bigcap_i \ker(U(Y_1 \times\cdots\times Y_k) \longrightarrow U(Y_1 \times\cdots\times \hat{Y}_i \times\cdots\times Y_k)).$$

The usefulness of these sequences is due to the fact that

<u>(i) these sequences are natural in the action $a: F \times E \to E$ as well as in U</u>, and

<u>(ii) if the action $a: F \times E \to E$ is principal</u> (i.e. $a(f,e) = e$ for any one $e \in E$ implies $f = 1$), <u>and</u>

$$U: \text{(pointed sets)} \longrightarrow \text{(groups)}$$

<u>then these sequences are exact</u>.

The first of these properties is obvious. To prove the second statement one uses the argument of [Kan-Whitehead, §16] and observes that the natural map $X \to X_{-1} \in \mathcal{S}$ is a homotopy equivalence and that therefore [Kan (HR)] the map $UX \to UX_{-1}$ is so too. As X_{-1} is discrete, so is UX_{-1} and thus $\pi_0 UX \approx UX_{-1}$, while $\pi_i UX = *$ for $i > 0$. This readily implies the exactness of the magic sequence.

The proof of the exactness of the other sequences is easy.

Now we turn to the

Proof of the principal fibration lemma. Consider the exact sequences of 3.1 with F and E as in 2.2 and $U = T^s R$, where T^s is as in the proof of Ch.I, 6.1. As an epimorphism of simplicial R-modules is a fibration with the kernel as fibre, the proof of Ch.I, 6.1 (R is solid) readily implies that

(*) all the simplicial R-modules in the exact sequences of 3.1 (and in particular the simplicial R-modules $U_k'X$) are s-connected.

Now start all over and let again F and E be as in 2.2, but put $U = \underset{\sim}{R}$. Then the sequences of 3.1 become exact sequences of group-like cosimplicial spaces (Ch.I, 2.2 and Ch.X, 4.8). As the functor Tot turns short exact sequences of group-like cosimplicial spaces into fibrations (Ch.X, 4.9 and 5.1), the finite product lemma (Ch.I, 7.2) implies that $\mathrm{Tot}\, U^k(F,\cdots,F,E)$ and $\mathrm{Tot}\, U^k(F,\cdots,F)$ and hence $\mathrm{Tot}\, U_k X$ are contractible for $k > 1$ and that the map $\mathrm{Tot}\, U_1 X \to \mathrm{Tot}\, UF$ is a homotopy equivalence. Furthermore the above observation (*) implies that for all k and s the fibre of the map $\mathrm{Tot}_{s+1}\, U_k' X \to \mathrm{Tot}_s\, U_k' X$ is connected and that therefore $\mathrm{Tot}\, U_k' X$ is connected for all k and in fact contractible for $k > 1$. Finally it is not hard to see that the map

$$R_\infty E \;=\; \mathrm{Tot}\, U_0 X \longrightarrow \mathrm{Tot}\, U_{-1} X \;=\; R_\infty B$$

is a fibration and that the inclusion of $R_\infty F = \mathrm{Tot}\, UF$ in its fibre is a homotopy equivalence.

We conclude by proving that $R_\infty E \to R_\infty B$ is, up to homotopy, a principal fibration. By the classification theorem [May, p.90] the map $E \to B$ fits in an induced fibre square

$$\begin{array}{ccc} E & \longrightarrow & WF \\ \downarrow & & \downarrow \\ B & \longrightarrow & \overline{W}F \end{array}$$

where $WF \to \overline{W}F$ is a principal fibration with WF contractible, and the desired result now follows from the fact that

$$\begin{array}{ccc} R_\infty E & \longrightarrow & R_\infty WF \\ \downarrow & & \downarrow \\ R_\infty B & \longrightarrow & R_\infty \overline{W}F \end{array}$$

is, up to homotopy, an induced fibre square, with $R_\infty WF$ contractible.

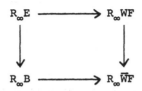

§4. Nilpotent fibrations

We will call a fibration $E \to B \in \mathcal{S}_{*c}$ with connected fibre F, nilpotent if "$\pi_1 E$ acts nilpotently on each $\pi_i F$". This turns out to be equivalent with requiring that the map $E \to B$ factors, up to homotopy, into a tower of principal fibrations with connected fibres that "get higher and higher connected". Corollary 2.4 of the principal fibration lemma thus implies that, up to homotopy, the R-completion preserves nilpotent fibrations.

We start with recalling

4.1 Nilpotent group actions. A group π acts on a group G if there is given a homomorphism

$$\alpha: \pi \longrightarrow \operatorname{Aut} G$$

and such an action is called nilpotent if there exists a finite sequence of subgroups of G

$$G = G_1 \supset \cdots \supset G_j \supset \cdots \supset G_n = *$$

such that for each j

(i) G_j is closed under the action of π,

(ii) G_{j+1} is normal in G_j and G_j/G_{j+1} is abelian, and

(iii) the induced action on G_j/G_{j+1} is trivial.

The notion of nilpotent action is a generalization of the notion of nilpotent group, as a group G is nilpotent if and only if the action of G on itself via inner automorphisms $((\alpha x)g = xgx^{-1}$ for all $x, g \in G)$ is nilpotent.

The following easy lemma will be needed:

4.2 Lemma. *If a group π acts on a short exact sequence of groups*

$$* \longrightarrow G' \longrightarrow G \longrightarrow G'' \longrightarrow * ,$$

then the action on G is nilpotent if and only if the actions on G' and G'' are so.

Now we define

4.3 Nilpotent fibrations. A *fibration* $p: E \to B \in \mathcal{J}_{*C}$ is called *nilpotent* if

(i) its fibre F is connected, and

(ii) the (obvious) action of $\pi_1 E$ on each $\pi_i F$ is nilpotent.

A *space* $X \in \mathcal{J}_{*C}$ is also called *nilpotent* if the action of $\pi_1 X$ on each $\pi_i X$ is nilpotent. Thus *a fibrant space $X \in \mathcal{J}_{*C}$ is nilpotent if and only if the fibration $X \to *$ is nilpotent.*

A useful property of nilpotent fibrations is

4.4 Proposition. *Let*

$$E_2 \xrightarrow{q} E_1 \xrightarrow{p} E_0 \qquad \in \mathcal{J}_{*C}$$

be two fibrations with connected fibres. If any two of p, q and pq are nilpotent fibrations, then so is the third.

Proof. If F_1, F_2 and F_{12} resp. denote the fibres of p, q and pq, then $\pi_1 E_2$ acts on the homotopy exact sequence

$$\cdots \longrightarrow \pi_{n+1}F_1 \longrightarrow \pi_n F_2 \longrightarrow \pi_n F_{12} \longrightarrow \pi_n F_1 \longrightarrow \cdots .$$

If $\pi_1 E_2$ acts nilpotently on any two of $\pi_* F_1$, $\pi_* F_2$ and $\pi_* F_{12}$, then it also acts nilpotently on the third (by 4.2). The proposition now follows readily.

4.5 Corollary. *Let* $q: E_2 \to E_1 \in \mathcal{S}_{*c}$ *be a fibration with connected fibre. If* E_1 *and* E_2 *are nilpotent spaces, then* q *is a nilpotent fibration.*

4.6 Corollary. *Let*

$$\cdots \longrightarrow E_n \longrightarrow \cdots \longrightarrow E_0 \qquad \in \mathcal{S}_{*c}$$

be a tower of principal fibrations with connected fibres that "get higher and higher connected". Then the composition

$$E_\infty = \lim_{\leftarrow} E_n \longrightarrow E_0 \qquad \in \mathcal{S}_{*c}$$

is a nilpotent fibration.

This corollary has a converse

4.7 Proposition. *Let* $p: E \to B \in \mathcal{S}_{*c}$ *be a nilpotent fibration. Then the Moore-Postnikov tower [May, p.34] of* p *can, up to homotopy, be refined to a tower of principal fibrations with connected fibres that "get higher and higher connected". In fact this can be done in such a manner that the fibres are* $K(\pi,n)$*'s [May, p.98].*

In view of this, corollary 2.4 of the principal fibration lemma can thus be restated as the

__4.8 Nilpotent fibration lemma.__ _Let_ $p\colon E \to B \in \mathcal{S}_{*C}$ _be a nilpotent fibration. Then_ $R_\infty p\colon R_\infty E \to R_\infty B$ _is also a nilpotent fibration and the inclusion_ $R_\infty(p^{-1}*) \to (R_\infty p)^{-1}*$ _is a homotopy equivalence._

Proof of 4.7. Let F be the fibre and let

$$\pi_i F = (\pi_i F)_1 \supset \cdots \supset (\pi_i F)_j \supset \cdots \supset (\pi_i F)_{n_i} = *$$

satisfy the conditions of 4.1 with respect to the action of $\pi_1 E$ on $\pi_i F$. Choose a strong deformation retract $E' \subset E$ for which the restriction $E' \to B$ is a _minimal_ fibration [May, p.140]. Then one can, for every pair of integers (i,j) with $1 \le j \le n_i$, construct a space $E^{(i,j)}$ by identifying two simplices $x, y \in E'_q$ whenever

(i) $px = py$,

(ii) the standard maps $\Delta x, \Delta y\colon \Delta[q] \to E'$ agree on the (i-1)-skeleton of $\Delta[q]$, and

(iii) the standard maps $\Delta x, \Delta y\colon \Delta[q] \to E'$ "differ" on every i-simplex of $\Delta[q]$ by an element of $(\pi_i F)_j$.

A straightforward calculation now yields that the $E^{(i,j)}$ form a tower of principal fibrations with the $K((\pi_i F)_j/(\pi_i F)_{j+1}, i)$ as fibres, which is a refinement of the Moore-Postnikov tower $\{E^{(i,1)}\}$ [May, p.34].

§5. The mod-R fibre lemma

We now come to the main result of this chapter, namely the

5.1 Mod-R fibre lemma. Let $p: E \to B \in \mathcal{S}_{*c}$ *be a fibration with connected fibre* F *and let the (Serre) action of* $\pi_1 B$ *on* $H_i(F; R)$ *be nilpotent for all* $i \geq 0$. *Then* $R_\infty p: R_\infty E \to R_\infty B$ *is a fibration and the inclusion*

$$R_\infty F = R_\infty(p^{-1}*) \longrightarrow (R_\infty p)^{-1}* \qquad \in \mathcal{S}_{*c}$$

is a homotopy equivalence.

This will be proved in §6 using the *nilpotent fibration lemma* 4.8 and the *fibre-wise R-completion* constructions of Ch.I, §8. A different proof will be given in Chapter III, §7.

In this section we shall show that 5.1 generalizes our previous fibre lemmas, and also applies to many new cases. We start with

5.2 Examples. The conditions of the mod-R fibre lemma are satisfied if, for instance

(i) $\pi_1 B = *$,

(ii) $E = F \times B$ *and* p *is the projection,*

(iii) the fibration $p: E \to B$ *is principal,*

(iv) $\pi_1 B$ *and* $H_i(F; R)$ *(*$i > 0$*) are all finite p-groups for p prime* (by [M. Hall, p.47] a finite p-group always acts nilpotently on another finite p-group).

A variation of the mod-R fibre lemma is the

5.3 Fibre square lemma. Let

$$\begin{array}{ccc} D & \longrightarrow & E \\ \downarrow & & \downarrow \\ A & \longrightarrow & B \end{array}$$

be a fibre square in $\mathcal{S}_{*c}$ *such that* $E \to B$ *satisfies the conditions of 5.1. Then*

$$\begin{array}{ccc} R_\infty D & \longrightarrow & R_\infty E \\ \downarrow & & \downarrow \\ R_\infty A & \longrightarrow & R_\infty B \end{array}$$

is, up to homotopy, a fibre square.

Proof. Apply 5.1 to both $D \to A$ and $E \to B$.

We conclude by deducing a

5.4 Mod-R nilpotent fibration lemma. Let $p\colon E \to B \in \mathcal{S}_{*c}$ *be a fibration with connected fibre* F *such that*

(i) $\pi_1 E$ *acts nilpotently on* $\pi_1 F$, *and*

(ii) $\pi_1 E$ *acts nilpotently on* $R \otimes \pi_i F$ *and* $\operatorname{Tor}(R, \pi_i F)$ *for each* $i > 1$.

Then the action of $\pi_1 B$ *on each* $H_i(F; R)$ *is nilpotent and hence (5.1)* $R_\infty p\colon R_\infty E \to R_\infty B$ *is a fibration and the inclusion*

$$R_\infty F = R_\infty(p^{-1}*) \longrightarrow (R_\infty p)^{-1}* \qquad \in \mathcal{S}_{*c}$$

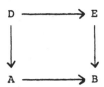

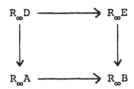

is a homotopy equivalence.

This result obviously applies to any nilpotent fibration and thus *the nilpotent fibration lemma 4.8 is indeed a special case of 5.1.*

Proof of 5.4. Apply the following lemmas (5.5 and 5.6) to the Moore-Postnikov tower [May, p.34] of $p: E \to B$.

5.5 Lemma. *Let*

$$E_2 \xrightarrow{q} E_1 \xrightarrow{p} E_0 \qquad\qquad \in \mathcal{S}_{*c}$$

be fibrations with connected fibres. If p and q satisfy the conditions of 5.1, then so does pq.

Proof. Let F_1, F_2 and F_{12} be the fibres of p, q and pq. Then the group $\pi_1 E_2$ acts on the mod-R homology (Serre) spectral sequence of the fibration

$$F_2 \longrightarrow F_{12} \longrightarrow F_1 .$$

To show that $\pi_1 E_2$ acts nilpotently on each $H_i(F_{12}, R)$ it will suffice (4.2) that $\pi_1 E_2$ acts nilpotently on each of the *twisted* homology groups

$$H_s(F_1; H_t(F_2; R)) \qquad\qquad \text{for} \quad s, t \geq 0.$$

Since $\pi_1 E_2$ acts nilpotently on the R-module $H_t(F_2; R)$ there is an R-module filtration

$$H_t(F_2; R) = \Gamma_1 \supset \cdots \supset \Gamma_j \supset \cdots \supset \Gamma_n = 0$$

such that each Γ_j is closed under the action of $\pi_1 E_2$ and each Γ_j/Γ_{j+1} has trivial $\pi_1 E_2$ action. Since $\pi_1 E_2$ acts nilpotently on each $H_i(F_1; R)$ it is now easy to show that $\pi_1 E_2$ also acts nilpotently on each $H_i(F_1; \Gamma_j/\Gamma_{j+1})$. This implies that $\pi_1 E_2$ acts nilpotently on $H_s(F_1; H_t(F_2; R))$ as required.

5.6 Lemma. Let $p: E \to B \in \mathcal{J}_{*C}$ be a fibration with $K(G,n)$ as fibre, such that, either

(i) $n = 1$ and $\pi_1 E$ acts nilpotently on G, or

(ii) $n \geq 2$ and $\pi_1 E$ acts nilpotently on $R \otimes G$ and $\mathrm{Tor}(R,G)$.

Then $\pi_1 B$ acts nilpotently on each $H_i(K(G,n); R)$.

Proof. Condition (i) makes the fibration nilpotent and the lemma then follows by combining 4.7 and 5.5. We now suppose (ii) and consider several cases.

The case $R \subset Q$. It is well-known that the obvious map $G \to R \otimes G$ induces an isomorphism

$$H_*(K(G,n); R) \approx H_*(K(R \otimes G,n); R)$$

and thus, by 2.7 and Ch.I, 5.5, a homotopy equivalence

$$R_\infty K(G,n) \simeq K(R \otimes G,n).$$

The desired result now follows easily from the fact that the fibration $\dot{R}_\infty E \to B$ (Ch.I, §8) is nilpotent.

The case $R = Z_p$. For p prime, the Cartan-Serre computations provide a natural isomorphism

$$H_*(K(G,n); Z_p) \approx U(QH_*(K(G,n); Z_p))$$

where $Q(-)$ is the indecomposable element functor and $U(-)$ is the homology version of the Steenrod-Epstein functor [see Bousfield-Kan (HS), 13.1], and, moreover, $QH_*(K(G,n); Z_p)$ is a natural direct sum of copies of $Z_p \otimes G$ and $Tor(Z_p, G)$. The desired result now follows using the analysis of $U(-)$ given in [Bousfield-Kan (HS), lemma 13.5].

<u>The case</u> $R = Z_{p^j}$. Using Bockstein exact sequences one can deduce this case from the case $R = Z_p$.

<u>The general case</u>. It suffices to show that $\pi_1 B$ acts nilpotently on each of the groups

$$H_i(K(G,n); R^t) \qquad\qquad H_i(K(G,n); R/R^t)$$

where R^t denotes the torsion subgroup of R. But by [Bousfield-Kan (CR), §3]

$$R/R^t \subset Q \qquad \text{and} \qquad R^t \approx \bigoplus_{p \in K} Z_{p^{e(p)}}$$

where K is a set of primes and each $e(p)$ is a positive integer and the desired result now follows from the previous special cases.

§6. Proof of the mod-R fibre lemma

We start with observing that, if, for a fibration $E \to B \in \mathcal{S}_{*c}$ with connected fibre F, the group $\pi_1 B$ acts nilpotently on each $H_i(F; R)$, then the tower of fibrations (Ch.I, §8)

$$\cdots \longrightarrow \dot{R}_s E \longrightarrow \dot{R}_{s-1} E \longrightarrow \cdots \longrightarrow \dot{R}_0 E = \dot{R}E \longrightarrow \dot{R}_{-1} E = B \quad \in \mathcal{S}_{*c}$$

is a tower of nilpotent fibrations. More precisely

6.1 Proposition. *Let $p\colon E \to B \in \mathcal{S}_{*c}$ be a fibration with connected fibre F. Then the action of $\pi_1 B$ on each $H_i(F; R)$ is nilpotent if and only if the fibration $\dot{R}p\colon \dot{R}E \to B$ (Ch.I, §8) is nilpotent.*

6.2 Proposition. *Let $p\colon E \to B \in \mathcal{S}_{*c}$ be a fibration with connected fibre F such that the action of $\pi_1 B$ on each $H_i(F; R)$ is nilpotent. Then the fibrations $\dot{R}_s E \to B$ (Ch.I, §8) are nilpotent for all $s < \infty$.*

Proof of 6.1. RF is the fibre of the fibration $\dot{R}p\colon \dot{R}E \to B$ and $\pi_1 RF$ acts trivially on each $\pi_i RF$.

Proof of 6.2. The fibre F_s of the fibration $\dot{R}_s E \to \dot{R}_{s-1} E$ is also the fibre of the fibration $R_s F \to R_{s-1} F$. Hence (2.6) F_s is connected and $\pi_1 R_s F$ acts trivially on the $\pi_i F_s$. The exactness of the sequence

$$\pi_1 R_s F \longrightarrow \pi_1 \dot{R}_s E \longrightarrow \pi_1 B \longrightarrow *$$

now implies that $\pi_1 B$ acts on each $\pi_i F_s$ (through $\pi_1 \dot{R}_s E$) and that it suffices to prove that these actions are nilpotent. But this is not hard to show using 4.2.

Now we turn to the

Proof of 5.1. Consider the commutative diagram

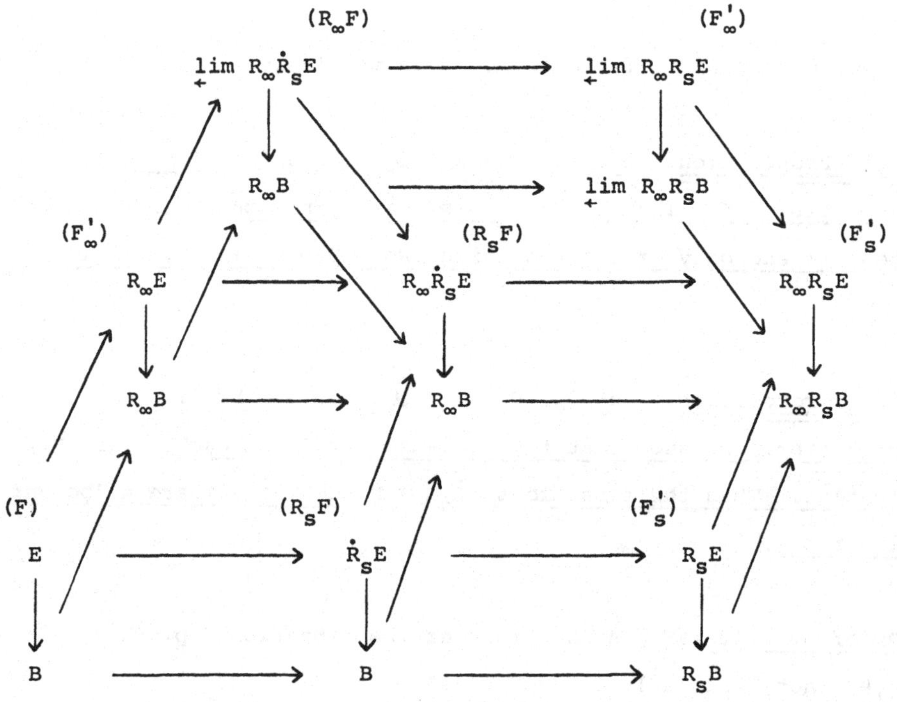

in which F'_s denotes the fibre of the fibration $R_s p\colon R_s E \to R_s B$ $(s \leq \infty)$ and the spaces in parentheses indicate either the actual fibre or a strong deformation retract thereof. It is not difficult to see (in view of 6.2, 4.8 and 2.7) that the fibres are indeed as indicated.

Using the triple lemma (Ch.I, 5.6) one readily shows that the composition

$$R_\infty E \longrightarrow \varprojlim R_\infty \dot{R}_s E \longrightarrow \varprojlim R_\infty R_s E$$

is the obvious map and thus a homotopy equivalence. This implies that *the map*

$$F'_\infty \longrightarrow \operatorname{fibre}(\varprojlim R_\infty \dot{R}_s E \longrightarrow R_\infty B)$$

induces a monomorphism on the homotopy groups.

Now form the analogue of the above diagram for the fibration $F \to *$ and map it into the above diagram. There results a commutative diagram

$$\begin{array}{ccc} \varprojlim R_\infty R_s F & \xrightarrow{\simeq} & \operatorname{fibre}(\varprojlim R_\infty \dot{R}_s E \longrightarrow R_\infty B) \\ \simeq \nearrow & & \nearrow \\ R_\infty F & \longrightarrow & F'_\infty \end{array} .$$

Clearly the indicated maps are homotopy equivalences and hence so is the map

$$R_\infty F = R_\infty(p^{-1}*) \longrightarrow (R_\infty p)^{-1}* = F'_\infty .$$

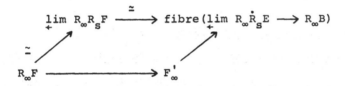

Chapter III. Tower lemmas

§1. Introduction

In this chapter we establish the following simple and useful sufficient conditions on a tower of fibrations $\{Y_s\}$, in order that it can be used to obtain the homotopy type of the R-completion of a given space X:

(i) If $f: X \to \{Y_s\}$ is a map which induces, for every R-module M, an isomorphism

$$\varinjlim H^*(Y_s; M) \approx H^*(X; M)$$

then f induces a homotopy equivalence $R_\infty X \simeq \varprojlim R_\infty Y_s$.

(ii) If, in addition, each Y_s is R-complete (Ch.I, 5.1), then the space $\varprojlim Y_s$ already has the same homotopy type as $R_\infty X$.

(iii) If, in addition, each Y_s satisfies the even stronger condition of being R-nilpotent (4.2), then, in a certain precise sense, the tower $\{Y_s\}$ has the same homotopy type as the tower $\{R_s X\}$.

We will actually formulate and prove these tower lemmas in terms of homology instead of cohomology, as this is not only more natural, but also easier, even though it requires a little bit of the pro-homotopy theory of [Artin-Mazur]. In more detail:

§2 We recall when a map between towers of groups is a pro-isomorphism (i.e. an isomorphism in the category of pro-groups) and show that these pro-isomorphisms behave essentially like ordinary isomorphisms; in particular they satisfy a five lemma.

<u>§3 and §4</u> Using these pro-isomorphisms we then define, for maps between towers of fibrations, a notion of <u>weak pro-homotopy equivalence</u>.

Examples of such weak pro-homotopy equivalences are all the various maps between towers of fibrations of Chapters I and II, which induce homotopy equivalences between the inverse limit spaces. These <u>tower versions</u> of the results of Chapters I and II are easily verified, except for the case of the mod-R fibre lemma (Ch.II, 5.1), which will be dealt with in §7.

<u>§5</u> contains a discussion of the notion of <u>R-nilpotent space</u>, i.e. a space for which the Postnikov tower can, up to homotopy, be refined to a tower of principal fibrations with simplicial R-modules as fibres. Examples are, for instance, all simplicial R-modules and the spaces $R_s X$ for $s < \infty$.

<u>§6</u> We state and prove the <u>tower lemmas</u> and show that, of course, the tower $\{R_s X\}$ satisfies the hypotheses of all three.

<u>§7</u> uses the strongest (R-nilpotent) tower lemma to prove <u>the tower version of the mod-R fibre lemma</u> (Ch.II, 5.1).

<u>§8</u> Here we interpret some of the preceding results to show that, up to homotopy, the R-completion of a space can be obtained in two steps:

(i) an <u>Artin-Mazur completion</u> yielding a "pro-homotopy type", followed by

(ii) a "<u>collapsing</u>" of the Artin-Mazur completion to an ordinary homotopy type.

This section is mainly intended for the categorically minded reader; we include a brief exposition of the relevant pro-category theory.

Notation. As in Chapter II we will mostly work in the category $\mathcal{S}_{*c}$ of pointed connected spaces. In view of the tower version (see 3.5) of the disjoint union lemma (Ch.I, 7.1) this is again no real restriction.

Of course (Ch.I, 4.5) we again assume throughout that the ring R is solid.

§2. Pro-isomorphisms of towers of groups

We recall from [Artin-Mazur] a few facts about pro-isomorphisms that are needed in this chapter, i.e. we

(i) explain when a map between towers of groups is a pro-isomorphism,

(ii) list various properties of pro-isomorphisms, and

(iii) observe that most of the results of this section also apply to pointed sets.

2.1 Pro-isomorphisms. A map $f: \{G_s\} \to \{H_s\}$ between two towers of groups (Ch.IX, 2.1) is called a pro-isomorphism if, for every group B, it induces an isomorphism

$$\varinjlim \operatorname{Hom}_{(\text{groups})}(H_s,B) \approx \varinjlim \operatorname{Hom}_{(\text{groups})}(G_s,B)$$

This is equivalent to the condition that, for every s, there is an integer $s' \geq s$ and a map $H_{s'} \to G_s$ such that the following diagram commutes

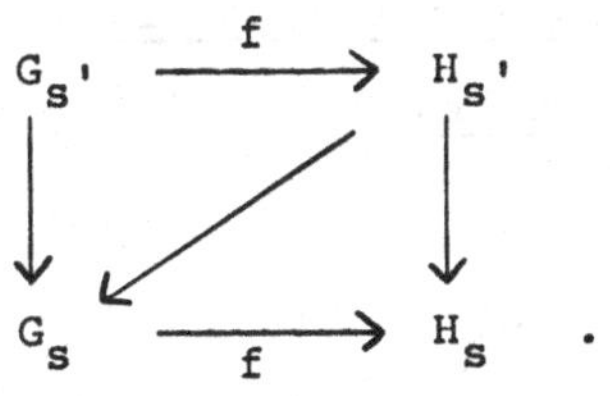

A tower of groups $\{K_s\}$ is called pro-trivial whenever the map $\{K_s\} \to \{*\}$ into the trivial tower is a pro-isomorphism.

Clearly the above definitions apply equally well to towers over an arbitrary pointed category, e.g. the category of pointed sets.

It is easy to show:

2.2 Proposition. A map of group towers $f: \{G_s\} \to \{H_s\}$ is a pro-isomorphism if and only if the pointed set towers $\{\ker f\}$ and $\{\operatorname{coker} f\}$ are pro-trivial, where $\{\operatorname{coker} f\}$ is formed by collapsing fG_s to a point.

Thus a map of group towers is a pro-isomorphism if and only if the underlying map of pointed set towers is a pro-isomorphism. Clearly 2.2 would remain valid if $\{\operatorname{coker} f\}$ were replaced by the tower of left cosets $\{H_s/fG_s\}$.

Some other immediate properties are:

2.3 Proposition. If $k \geq 0$ and $\{G_s\}$ is a tower of groups, then the inclusion of the k-th derived tower (Ch.IX, 2.2) $\{G_s^{(k)}\} \to \{G_s\}$ is a pro-isomorphism.

This also holds for pointed set towers.

2.4 Proposition. Let $f: \{G_s\} \to \{H_s\}$ and $g: \{H_s\} \to \{K_s\}$ be maps of group towers. If any two of the maps f, g and gf are pro-isomorphisms, then so is the third.

This also holds for pointed set towers.

2.5 Proposition. Let

$$\{G_s\} \longrightarrow \{H_s\} \longrightarrow \{K_s\}$$

be an exact sequence of maps of group towers. If $\{G_s\}$ and $\{K_s\}$ are pro-trivial, then so is $\{H_s\}$.

This also holds for <u>pointed set</u> towers.

<u>2.6 Proposition</u>. <u>Let</u> $f: \{G_s\} \to \{H_s\}$ <u>be a map of group towers which is a pro-isomorphism. Then</u> f <u>induces isomorphisms (Ch.IX, 2.1)</u>

$$\varprojlim G_s \approx \varprojlim H_s \qquad \text{and} \qquad \varprojlim{}^1 G_s \approx \varprojlim{}^1 H_s .$$

Of course, the first part also holds for <u>pointed set</u> towers.

<u>Proof of 2.6</u>. Obtain a tower

$$\cdots \longrightarrow G_{i_{s+1}} \xrightarrow{f} H_{i_{s+1}} \longrightarrow G_{i_s} \xrightarrow{f} H_{i_s} \longrightarrow \cdots$$

by interweaving a cofinal subtower of $\{G_s\}$ with a cofinal subtower of $\{H_s\}$ and then apply (Ch.IX, 3.1) to the corresponding tower of fibrations of $K(\pi,1)$'s.

Using 2.2 and a large amount of diagram chasing one can also obtain a

<u>2.7 Five lemma</u>. <u>Let</u>

$$\begin{array}{ccccccccc}
\{G_s\} & \longrightarrow & \{H_s\} & \longrightarrow & \{K_s\} & \longrightarrow & \{L_s\} & \longrightarrow & \{M_s\} \\
\downarrow{\scriptstyle g} & & \downarrow{\scriptstyle h} & & \downarrow{\scriptstyle k} & & \downarrow{\scriptstyle l} & & \downarrow{\scriptstyle m} \\
\{G'_s\} & \longrightarrow & \{H'_s\} & \longrightarrow & \{K'_s\} & \longrightarrow & \{L'_s\} & \longrightarrow & \{M'_s\}
\end{array}$$

<u>be a diagram of group towers in which both rows are exact, the maps</u> h <u>and</u> l <u>are pro-isomorphisms and</u> $\{\text{coker } g\}$ <u>and</u> $\{\ker m\}$ <u>are pro-trivial. Then the map</u> k <u>is also a pro-isomorphism</u>.

§3. Weak pro-homotopy equivalences

We now consider, for maps between towers of fibrations, a notion of weak pro-homotopy equivalence, and observe that many of the homotopy equivalences of Chapters I and II are induced by such weak pro-homotopy equivalences.

3.1 Weak pro-homotopy equivalences. A map $\{X_s\} \to \{Y_s\}$ between towers of fibrations in $\mathcal{S}_{*c}$ will be called a weak pro-homotopy equivalence if the induced maps

$$\{\pi_i X_s\} \longrightarrow \{\pi_i Y_s\} \qquad i \geq 1$$

are pro-isomorphisms. This corresponds to the notion of #-isomorphism of [Artin-Mazur, §4].

Clearly 2.6 and Ch.IX, 3.1, imply that every weak pro-homotopy equivalence $\{X_s\} \simeq \{Y_s\}$ induces a homotopy equivalence $\varprojlim X_s \simeq \varprojlim Y_s$.

The following propositions of [Artin-Mazur, §4] indicate that the term "weak pro-homotopy equivalence" is indeed an appropriate one.

3.2 Proposition. For every tower of fibrations $\{X_s\}$ in $\mathcal{S}_{*c}$, the natural map into its "Postnikov tower" [May, p.31]

$$\{X_s\} \longrightarrow \{X_s^{(s)}\}$$

is a weak pro-homotopy equivalence.

The proof is trivial.

3.3 Proposition. *A map $\{X_s\} \to \{Y_s\}$ between two towers of fibrations in $\mathcal{S}_{*c}$ is a weak pro-homotopy equivalence if and only if the induced map (see 3.2) $\{X_s^{(s)}\} \to \{Y_s^{(s)}\}$ of their "Postnikov towers" is a pro-isomorphism of towers over the pointed homotopy category (see 2.1)*, i.e. if and only if, for every s, there is an integer $s' > s$ and a map $Y_{s'}^{(s')} \to X_s^{(s)}$ such that in the following diagram both triangles commute up to homotopy

$$\begin{array}{ccc} X_{s'}^{(s')} & \longrightarrow & Y_{s'}^{(s')} \\ \downarrow & \swarrow & \downarrow \\ X_s^{(s)} & \longrightarrow & Y_s^{(s)} \end{array} .$$

The proof is rather long and will be postponed until §4.

For future reference we note:

3.4 Corollary. *A weak pro-homotopy equivalence $\{X_s\} \to \{Y_s\}$ between towers of fibrations in $\mathcal{S}_{*c}$ induces, for every abelian group G, pro-isomorphisms*

$$\{H_n(X_s;\ G)\} \simeq \{H_n(Y_s;\ G)\} \qquad n \geq 0.$$

We end with some remarks on the

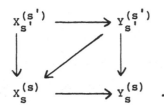

3.5 Tower versions of previous results. Many of the homotopy equivalences obtained in Chapters I and II have *tower versions*, i.e. they are induced by a weak pro-homotopy equivalence between towers of fibrations. This is very easy to verify for the results of Chapter I. The tower version of the mod-R fibre lemma (Ch.II, 5.1) will be proved in §7 with the use of the following tower version of the principal fibration lemma (Ch.II, 2.2).

3.6 Tower version of the principal fibration lemma. Let $p: E \to B \in \mathcal{S}_{*c}$ be a principal fibration with connected fibre F. Then the induced map of towers of fibrations

$$\{R_sF\} \longrightarrow \{F'_s\},$$

where F'_s denotes the fibre of $R_sp: R_sE \to R_sB$, is a weak pro-homotopy equivalence.

Proof. Using Ch.X, 4.9, one can show that the maps $F'_s \to F'_{s-1}$ are indeed fibrations, while Ch.I, 6.2 implies that the F'_s are connected.

We now use the notation of the last part of Ch.II, §3 and observe that the results of §2 readily imply that the towers

$$\{\pi_i \operatorname{Tot}_s U^k(F,\cdots,F,E)\} \qquad \text{and} \qquad \{\pi_i \operatorname{Tot}_s U^k(F,\cdots,F)\}$$

are pro-trivial for $k > 1$. Hence the towers $\{\pi_i \operatorname{Tot}_s U_kX\}$ are pro-trivial for $k > 1$ and the towers $\{\pi_i \operatorname{Tot}_s U'_kX\}$ are so for $k > 0$. The desired result now follows from the fact that the maps of towers

$$\{\pi_i \operatorname{Tot}_s U_1X\} \longrightarrow \{\pi_i \operatorname{Tot}_s UF\} = \{\pi_i R_sF\}$$

$$\{\pi_i \operatorname{Tot}_s U_1X\} \longrightarrow \{\pi_i \operatorname{Tot}_s U'_0X\}$$

are pro-isomorphisms.

§4. Proof of 3.3

The "if" part of 3.3 is trivial.

To prove the "only if" part we first need

4.1 Lemma. *Let $\{X_s\}$ be a tower of fibrations in $\mathcal{S}_*$ such that $\{\pi_n X_s\}$ is pro-trivial for each $n \geq 0$. Then, for each s and t, there exists an integer q (which depends on s and t) such that, up to homotopy, the map $X_{s+q} \to X_s$ factors through the Eilenberg subspace $E^t X_s \subset X_s$* ($E^t X_s$ is the fibre of the Postnikov map $X_s \to X_s^{(t)}$ [May, p.31]).

Proof. The lemma is clear for the tower $\{E^k X_s\}$, when $k = t-1$; and this easily implies the general case.

We also need

4.2 Lemma. *Let $\{X_s\} \to \{Y_s\}$ be a weak pro-homotopy equivalence between towers of fibrations in $\mathcal{S}_{*c}$, and let N be a $\pi_1 Y_k$-module for some $k \geq 0$. Then the induced map*

$$\varinjlim H^*(Y_s;\ N) \longrightarrow \varinjlim H^*(X_s;\ N) \qquad \text{(twisted coefficients)}$$

is an isomorphism.

Proof. We may suppose that each map $X_s \to Y_s$ is a fibration with fibre F_s. By 2.7 (slightly modified for $n = 0$) the towers $\{\pi_n F_s\}$ $(n \geq 0)$ then are pro-trivial. Thus, by 4.1, the direct limit of the E_2-terms of the Serre spectral sequences of the fibrations $X_s \to Y_s$ satisfies

$$\varinjlim H^*(Y_s; H^*(F_s; N)) \approx \varinjlim H^*(Y_s; N) \quad \text{(twisted coefficients)}$$

and the lemma now follows from the fact that the direct limit Serre spectral sequence converges to $\varinjlim H^*(X_s; N)$.

Finally we prove the following lemma which readily implies the "only if" part of 3.3.

4.3 Lemma. Let $\{X_s\} \to \{Y_s\}$ be a map of towers of fibrations in $\mathcal{S}_{*c}$ such that $\{\pi_1 X_s\} \to \{\pi_1 Y_s\}$ is a pro-isomorphism and

$$\varinjlim H^*(Y_s; N) \approx \varinjlim H^*(X_s; N) \quad \text{(twisted coefficients)}$$

for each $\pi_1 Y_k$-module N with $k \geq 0$. Then, for every space $V \in \mathcal{S}_{*c}$ which is fibrant (i.e. $V \to *$ is a fibration) the induced maps between the direct limits of the pointed homotopy classes of maps

$$\varinjlim [Y_s, V^{(n)}] \longrightarrow \varinjlim [X_s, V^{(n)}] \qquad 0 \leq n < \infty$$

are isomorphisms.

Proof. We may suppose that each map $X_s \to Y_s$ is an inclusion. To prove that

$$\varinjlim [Y_s, V^{(n)}] \longrightarrow \varinjlim [X_s, V^{(n)}] \qquad 0 \leq n < \infty$$

is onto, it suffices to show that, for each commutative square

$$\begin{array}{ccc} X_s & \longrightarrow & V^{(n)} \\ \downarrow & & \downarrow \\ Y_s & \longrightarrow & V^{(n-1)} \end{array}$$

there exists a $q \geq s$ such that the square

$$\begin{array}{ccc} X_q & \longrightarrow & V^{(n)} \\ \downarrow & \overset{u}{\nearrow} & \downarrow \\ Y_q & \longrightarrow & V^{(n-1)} \end{array}$$

has a map u making both triangles commute. For $n > 1$ the obstruction to finding u lies in

$$H^{n+1}(Y_q, X_q; \pi_n V) \qquad \text{(twisted cohomology)}$$

and for $n = 1$ the obstruction is expressed by

$$\begin{array}{ccc} \pi_1 X_q & \longrightarrow & \pi_1 V \\ \downarrow & \nearrow & \\ \pi_1 Y_q & & \end{array}$$

In both cases the obstruction can be killed by taking q large enough. The 1-1 part of the lemma can be proved similarly, or, alternatively, can be deduced from the onto part using the inclusion

$$\{(\dot{\Delta}[1] \times Y_s) \cup (\Delta[1] \times X_s)\} \longrightarrow \{\Delta[1] \times Y_s\}.$$

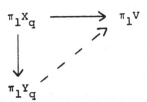

§5. R-nilpotent spaces

In this section we discuss the notion of R-nilpotent spaces, i.e. spaces for which the Postnikov tower can, up to homotopy, be refined to a tower of principal fibrations with simplicial R-modules as fibres. It turns out that the R-nilpotent spaces are exactly the spaces for which the natural map

$$\{X\} \longrightarrow \{R_s X\}$$

is a weak pro-homotopy equivalence. Useful examples are simplicial R-modules and the spaces $R_s X$ for $s < \infty$.

We start with defining:

5.1 R-nilpotent groups. A group G is said to be R-nilpotent if it has a finite central series

$$G = G_1 \supset \cdots \supset G_j \supset \cdots \supset G_n = *$$

such that each quotient G_j/G_{j+1} admits an R-module structure (which by [Bousfield-Kan (CR), 2.5] is unique.

Clearly a Z-nilpotent group is the same as a nilpotent group, and more generally (see Ch.V, 2.6), a $Z[J^{-1}]$-nilpotent group is the same as a uniquely J-divisible nilpotent group. It is also evident that a Z_p-nilpotent group is the same as a nilpotent group in which the order of each element divides p^k for some fixed $k < \infty$.

5.2 R-nilpotent spaces. A space $X \in \mathcal{S}_{*c}$ will be called R-nilpotent if

(i) X is nilpotent (Ch.II, 4.3), and

(ii) $\pi_i X$ is R-nilpotent for each $i \geq 1$.

An obvious example of an R-nilpotent space is any *connected simplicial R-module*. And clearly *a Z-nilpotent space is the same as a nilpotent space* (Ch.II, 4.3).

Now we state the main result of this section.

5.3 Proposition. *For a space* $X \in \mathcal{S}_{*C}$, *the following three conditions are equivalent:*

(i) X is R-nilpotent.

(ii) The natural map of towers

$$\{X\} \longrightarrow \{R_s X\}$$

is a weak pro-homotopy equivalence.

(iii) The Postnikov tower of X [May, p.31] can, up to homotopy, be refined to a tower of principal fibrations with as fibres $K(\pi,n)$*'s for which* $n \geq 1$ *and* π *admits an R-module structure.*

5.4 Corollary.

(i) If $X \in \mathcal{S}_{*C}$ *is R-nilpotent, then X is R-complete*, i.e. the map $\phi\colon X \to R_\infty X$ is a weak equivalence.

(ii) If $\{X_s\}$ *is a tower of fibrations in* $\mathcal{S}_{*C}$ *such that each* X_s *is R-nilpotent, then the map*

$$\phi\colon \{X_s\} \longrightarrow \{R_s X_s\}$$

is a weak pro-homotopy equivalence.

Proof of 5.3. *(i) → (iii)* is proved in the same way as Ch.II,

4.7, using 5.7 below.

(iii) → (ii) is proved by combining 2.7, 3.6, Ch.II, 2.2 and Ch.II, 2.7.

(ii) → (i). It follows from 3.3 that (ii) implies that each $X^{(s)}$ is, up to homotopy, a retract of $(R_k X)^{(s)}$ for some k (which depends on s). Using 5.6 and 5.8 below and Ch.II, 4.2, it is now not hard to show that each $X^{(s)}$ is R-nilpotent. This implies (i).

5.5 Proposition. *Let $p: E \to B \in \mathcal{S}_{*c}$ be a principal fibration with connected fibre F. If any two of F, E and B are R-nilpotent, then so is the third.*

This follows easily from 5.8 below and Ch.II, 4.2.

Combining 5.5 with Ch.II, 2.6 one gets:

5.6 Corollary. *Let $X \in \mathcal{S}_{*c}$. Then $R_s X$ is R-nilpotent for all $s < \infty$.*

5.7 Lemma. *Let G be an R-nilpotent abelian group on which a group π acts nilpotently. Then there is a finite sequence of subgroups of G*

$$G = G_1 \supset \cdots \supset G_j \supset \cdots \supset G_n = *$$

such that for each j

(i) G_j is closed under the action of π,

(ii) the induced action on G_j/G_{j+1} is trivial, and

(iii) the quotient G_j/G_{j+1} admits an R-module structure.

Proof. The "center of G under the action of π", i.e. the group

$$\{g \in G | \; xg = g \quad \text{for all} \quad x \in \pi\}$$

is R-nilpotent by 5.8 below, because it is the kernel of a homomorphism from G to a product of copies of G. The desired filtration of G can thus be obtained by taking the "upper central series of G under the action of π".

5.8 Lemma. Let $f: G \to H$ be a homomorphism between R-nilpotent groups. Then $\ker f$ is R-nilpotent; and if the image of f is normal in H, then $\operatorname{coker} f$ is also R-nilpotent.

Proof. If G and H are R-modules, then $\ker f$ and $\operatorname{coker} f$ admit R-module structures since f is necessarily R-linear [Bousfield-Kan (CR), 2.4].

In the general case choose central series

$$G = G_1 \supset \cdots \supset G_j \supset \cdots \supset G_n = *$$

$$H = H_1 \supset \cdots \supset H_j \supset \cdots \supset H_n = *$$

such that for each j

(i) $fG_j \subset H_j$ and

(ii) G_j/G_{j+1} and H_j/H_{j+1} admit R-module structures.

(The desired pair of central series can be obtained by reindexing an arbitrary pair). The map f induces additive relations [MacLane, p.51]

$$d_s\colon G_j/G_{j+1} \longrightarrow H_{j+s}/H_{j+s+1}$$

given by $d_s[x] = [fx]$ for each $x \in G_j$ with $fx \in H_{j+s}$. Using these relations one obtains a spectral sequence of R-modules. Passing to the E_∞-term, one gets that the abelian groups

$$(G_{j+1}(\ker f \cap G_j))/G_{j+1}$$

$$H_j/(H_{j+1}(\mathrm{im}\, f \cap H_{j+1}))$$

admit R-module structures. But these abelian groups are precisely the quotients of the central series

$$\{\ker f \cap G_j\} \qquad \text{for } \ker f$$

$$\{\mathrm{im}\,(H_j \longrightarrow \mathrm{coker}\, f)\} \qquad \text{for } \mathrm{coker}\, f \text{ (if it exists)}$$

This proves the lemma.

§6. The tower lemmas

To simplify the formulation of the tower lemmas we define a notion of

6.1 R-towers for a space X. By an R-tower for a space $X \in \mathcal{S}_{*C}$ we mean a tower of fibrations $\{Y_s\}$ in $\mathcal{S}_{*C}$ together with a map $\{X\} \to \{Y_s\}$ which induces, for every $i \geq 1$, a pro-isomorphism

$$\{H_i(X; R)\} \approx \{H_i(Y_s; R)\}$$

or equivalently (see 6.7), for every R-module M, an isomorphism

$$\varinjlim H^*(Y_s; M) \approx H^*(X; R).$$

Then one has the main

6.2 Tower lemma. Let $X \in \mathcal{S}_{*C}$ and let $f: \{X\} \to \{Y_s\}$ be an R-tower for X. Then f induces a weak pro-homotopy equivalence

$$\{R_sX\} \simeq \{R_sY_s\}$$

and hence a homotopy equivalence

$$R_\infty X = \varprojlim R_sX \simeq \varprojlim R_sY_s = \varprojlim R_\infty Y_s .$$

This result can be strengthened by requiring that each Y_s is R-complete or even R-nilpotent:

6.3 R-complete tower lemma. Let $X \in \mathcal{S}_{*C}$ and let $f: \{X\} \to \{Y_s\}$ be an R-tower for X such that each Y_s is R-complete.

Then the induced map

$$\lim_{\leftarrow} Y_s \longrightarrow \lim_{\leftarrow} R_s Y_s = \lim_{\leftarrow} R_\infty Y_s$$

is a homotopy equivalence and hence $\lim_{\leftarrow} Y_s$ *already has the same homotopy type as* $R_\infty X$.

6.4 R-nilpotent tower lemma. *Let* $X \in \mathcal{S}_{*C}$ *and let* $f: \{X\} \to \{Y_s\}$ *be an R-tower for* X *such that each* Y_s *is R-nilpotent. Then the towers* $\{Y_s\}$ *and* $\{R_s X\}$ *have the same "weak pro-homotopy type".*

As one might expect, *for every* $X \in \mathcal{S}_{*C}$, *the natural map* $\{X\} \to \{R_s X\}$ *satisfies the conditions of all three tower lemmas*. This follows immediately from 5.6 and the following result of [Dror (C)], which originally suggested the existence of the tower lemmas.

6.5 Proposition. *For every* $X \in \mathcal{S}_{*C}$, *the natural map* $\{X\} \to \{R_s X\}$ *is an R-tower for* X.

The above results (6.2-6.5) are easy consequences of 5.4, the triple lemma (Ch.I, 5.6) and

6.6 Proposition. *Let* $\{X_s\} \to \{Y_s\}$ *be a map of towers of fibrations in* $\mathcal{S}_{*C}$. *Then the induced map*

$$\{R_s X_s\} \longrightarrow \{R_s Y_s\}$$

is a weak pro-homotopy equivalence if and only if the induced map

$$\{RX_s\} \longrightarrow \{RY_s\}$$

is a weak pro-homotopy equivalence, i.e. if and only if, for every integer $i \geq 1$, the induced map

$$\{H_i(X_s; R)\} \longrightarrow \{H_i(Y_s; R)\}$$

is a pro-isomorphism.

Proof. To prove the "if" part, observe that 3.4 implies that the maps $\{R^n X_s\} \to \{R^n Y_s\}$ are weak pro-homotopy equivalences for all $n \geq 1$. The desired result now is not hard to prove, using 2.7, Ch.I, 6.1 and Ch.X, 6.3.

The "only if" part is an easy consequence of 3.4 and the fact that $\phi_*: H_*(X; R) \to H_*(R_s X; R)$ has a natural left inverse (Ch.I, 5.4).

We end this section with a result which may help clarify the notion of R-tower.

6.7 Proposition. Let $X \in \mathcal{S}_{*c}$ and let $\{Y_s\}$ be a tower of fibrations in $\mathcal{S}_{*c}$. For a map $\{X\} \to \{Y_s\}$ the following four conditions then are equivalent:

(i) $\{X\} \to \{Y_s\}$ is an R-tower for X.

(ii) For every injective R-module I

$$\varinjlim H^*(Y_s; I) \approx H^*(X; I)$$

(iii) For every R-module M

$$\varinjlim H^*(Y_s; M) \approx H^*(X; M)$$

(iv) For every R-nilpotent space $V \in \mathcal{S}_{*c}$ which is fibrant

<u>(i.e. $V \to *$ is a fibration)</u>

$$\lim_{\to} [Y_s, V^{(n)}] \approx [X, V^{(n)}] \qquad 0 \le n < \infty .$$

<u>Proof</u>. <u>(i) ⟺ (ii)</u>. This follows from the fact that

$$H^*(X; I) \approx \mathrm{Hom}_R(H_*(X; R), I)$$

<u>(ii) ⟹ (iii)</u>. Suppose (ii) and let I^* be an injective resolution for M. Then for each t

$$\lim_{\to} \mathrm{Hom}_R(H_t(Y_s; R), I^*) \approx \mathrm{Hom}_R(H_t(X; R), I^*)$$

and hence, for $n \ge 0$,

$$\lim_{\to} \mathrm{Ext}^n_R(H_*(Y_s; R), M) \approx \mathrm{Ext}^n_R(H_*(X; R), M).$$

Now (iii) can be deduced using the universal coefficient spectral sequence.

<u>(iii) ⟹ (ii)</u>. This is easy.

<u>(iii) ⟹ (iv)</u>. Suppose (iii) and express the Postnikov space $V^{(n)}$ as the inverse limit of a finite tower of principal fibrations with as fibres $K(\pi,n)$'s for which $n \ge 1$ and π admits an R-module structure (5.3). Then (iv) follows by an untwisted version of the argument used to prove 4.3.

<u>(iv) ⟹ (iii)</u>. This again is easy.

§7. Tower version of the mod-R fibre lemma

As promised in 3.5, we now prove the

7.1 Tower version of the mod-R fibre lemma. Let $E \to B \in \mathcal{S}_{*c}$ be a fibration with connected fibre F and let the (Serre) action of $\pi_1 B$ on each $H_i(F; R)$ be nilpotent. Then the induced map of towers of fibrations

$$\{R_s F\} \longrightarrow \{F_s'\}$$

where F_s' denotes the fibre of the map $R_s E \to R_s B$, is a weak pro-homotopy equivalence.

This result easily implies the mod-R fibre lemma (Ch.II, 5.1), and thus our proof below can be used in place of the earlier proof (Ch.II, §6).

Proof (not using Ch.II, 5.1). Using Ch.X, 4.9, one can show that the maps $F_s' \to F_{s-1}'$ are fibrations, while Ch.I, 6.2 implies that the F_s' are connected. Furthermore it is not hard to show, using 5.7, 5.8 and Ch.II, 4.2, that the F_s' are R-nilpotent. By the R-nilpotent tower lemma (6.4) it thus remains to show that the natural map $\{F\} \to \{F_s'\}$ is an R-tower for F.

To do this we consider the (obvious) map from "the Serre spectral sequence of the fibration $E \to B$" to "the tower of Serre spectral sequences of the fibrations $R_s E \to R_s B$" and show, by induction on k that

$(i)_k$ this spectral sequence map is a pro-isomorphism on $E^2_{p,k}$ for all p, and

$(ii)_k$ <u>the map</u> $\{H_k(F;\ R)\} \to \{H_k(F'_s;\ R)\}$ <u>is a pro-isomorphism.</u>

For this we need the following tower version of the Zeeman comparison theorem, c.f. [Quillen (PG), 3.8]:

<u>7.2 Tower comparison lemma.</u> <u>Let</u>

$$\{E^2_{p,q}(X_s) \Longrightarrow H_{p+q}(X_s)\} \xrightarrow{f} \{E^2_{p,q}(Y_s) \Longrightarrow H_{p+q}(Y_s)\}$$

<u>be a map of towers of first quadrant spectral sequences of homological type. If</u> $H_n(f)$ <u>is a pro-isomorphism for all</u> n <u>and</u> $E^2_{p,q}(f)$ <u>is a pro-isomorphism for</u> $q < k$, <u>then</u>

<u>(i)</u> $E^2_{0,k}(f)$ <u>is a pro-isomorphism, and</u>

<u>(ii)</u> $E^2_{1,k}(f)$ <u>is a pro-epimorphism</u> (i.e. the cokernel is pro-trivial).

<u>Continuation of the proof of 7.1.</u> Clearly (6.5) $(i)_0$ and $(ii)_0$ hold. Now assume $(i)_j$ and $(ii)_j$ for $0 \le j < k$. Then (6.5) the map $\{H_i(E;\ R)\} \to \{H_i(R_sE;\ R)\}$ is a pro-isomorphism for all $i \ge 0$ and hence 7.2 implies that <u>the spectral sequence map is a pro-isomorphism on</u> $E^2_{0,k}$ <u>and a pro-epimorphism on</u> $E^2_{1,k}$.

Next put

$$M = H_k(F;\ R) \qquad \text{and} \qquad M_s = H_k(F'_s;\ R)$$

and let

$$I \subset Z\pi_1B \qquad \text{and} \qquad I_s \subset Z\pi_1R_sB$$

denote the augmentation ideals of the group rings. As π_1B <u>acts nilpotently on each</u> $H_i(F;\ R)$ and as (in view of 5.6 and Ch.II, 4.5 and 5.4) π_1R_sB <u>acts nilpotently on each</u> $H_i(F'_s;\ R)$, it is not hard to see that, in order to obtain $(i)_k$ and $(ii)_k$, it suffices to show that for all $n \geq 1$

(iii)$_n$ <u>the maps</u>

$$\{H_p(B;\ M/I^nM)\} \longrightarrow \{H_p(R_sB;\ M_s/I_s^nM_s)\} \qquad p \geq 0$$

$$\{M/I^nM\} \longrightarrow \{M_s/I_s^nM_s\}$$

<u>are all pro-isomorphisms</u>.

This is again done by induction. The case $n = 1$ is clear since

$$H_0(B;\ M) \approx M/IM \qquad \text{and} \qquad H_0(R_sB;\ M_s) \approx M_s/I_sM_s \ .$$

For the induction step one considers the map from the exact sequence

$$\cdots \longrightarrow H_1(B;\ M) \longrightarrow H_1(B;\ M/I^nM) \longrightarrow H_0(B;\ I^nM) \longrightarrow$$
$$\longrightarrow H_0(B;\ M) \longrightarrow H_0(B;\ M/I^nM) \longrightarrow *$$

to the tower of exact sequence involving the R_sB and M_s. Using the five lemma (2.7) one gets that the map

$$\{I^nM/I^{n+1}M\} \approx \{H_0(B;\ I^nM)\} \longrightarrow \{H_0(R_sB;\ I_s^nM_s)\} \approx \{I_s^nM_s/I_s^{n+1}M_s\}$$

is a pro-isomorphism and from this one readily obtains (iii)$_{n+1}$.

§8. An Artin-Mazur-like interpretation of the R-completion

We first give a brief general exposition of Artin-Mazur completions [Artin-Mazur, §3], and we then show that, up to homotopy, the R-completion of a space may be obtained in two steps:

(i) an Artin-Mazur completion yielding a "pro-homotopy type", followed by

(ii) a "collapsing" of the Artin-Mazur completion to an ordinary homotopy type.

The proof is based on the observation that, roughly speaking, $\{R_s X\}$ is cofinal in the system of R-nilpotent target spaces of X (see 8.3).

This Artin-Mazur-like interpretation may help to clarify and justify the R-completion; however, the reader interested in "real mathematics" may safely skip this section. A different Artin-Mazur-like interpretation of the R-completion is given in Chapter XI.

To explain Artin-Mazur completions we need

8.1 Categories of pro-objects [Artin-Mazur, Appendix]. Let $\mathcal{C}$ be a category. A pro-object over $\mathcal{C}$ then is an I-diagram (Ch.XI, 3.1) over $\mathcal{C}$ where I is a small left filtering (Ch.XI, 9.3). The pro-objects over $\mathcal{C}$ form a category pro-$\mathcal{C}$ with maps defined by

$$\operatorname{Hom}_{\text{pro-}\mathcal{C}}(\underline{X}, \underline{Y}) = \varprojlim_j (\varinjlim_i \operatorname{Hom}_{\mathcal{C}}(\underline{X}_i, \underline{Y}_j))$$

where $\underline{X} = \{X_i\}_{i \in I}$ and $\underline{Y} = \{Y_j\}_{j \in J}$.

Clearly $\mathcal{C}$ is a full subcategory of pro-$\mathcal{C}$, and any functor $T\colon \mathcal{C} \to \mathcal{D}$ prolongs in an obvious way to a functor

$$\text{pro-}T\colon \text{pro-}\mathcal{C} \longrightarrow \text{pro-}\mathcal{D}.$$

Note that in §2 we were "really" working in the category pro-(*groups*); for instance, a map of group towers is a pro-isomorphism (2.1) if and only if it corresponds to an isomorphism in the category pro-(*groups*).

We can now introduce a categorical version of

8.2 The Artin-Mazur completion [Artin-Mazur, §3]. Let $\mathcal{M}$ be a full subcategory of a category $\mathcal{C}$, with inclusion functor $\mu:\mathcal{M}\to\mathcal{C}$; and for each object $X \in \mathcal{C}$, let $\mu\backslash X$ denote the category whose objects are maps $X \to M \in \mathcal{C}$ with $M \in \mathcal{M}$, and whose maps are the commutative triangles

$$\begin{array}{ccc} & & M \\ & \nearrow & \downarrow \\ X & & \\ & \searrow & \downarrow \\ & & M' \end{array} \qquad \in \mathcal{C} \quad .$$

Then pro-$\mathcal{M}$ is a full subcategory of pro-$\mathcal{C}$ and one has:

If, for each $X \in \mathcal{C}$, there exists a left cofinal (Ch.XI, 9.3) functor $I \to \mu\backslash X$, where I is a small left filtering, then the inclusion functor

$$\text{pro-}\mu\colon\ \text{pro-}\mathcal{M} \longrightarrow \text{pro-}\mathcal{C}$$

has a left adjoint

$$U\colon\ \text{pro-}\mathcal{C} \longrightarrow \text{pro-}\mathcal{M} \quad .$$

Thus, for $\underline{X} \in$ pro-$\mathcal{C}$, the adjunction map $\underline{X} \to U\underline{X} \in$ pro-$\mathcal{C}$ is the *universal example* of a map from $\underline{X}$ to an object of pro-$\mathcal{M}$; and we shall call $U\underline{X}$ the *Artin-Mazur completion* of $\underline{X}$.

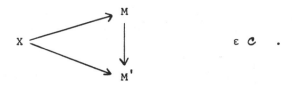

For $X \in \mathcal{C}$, it can be shown that, as one might expect, $UX \in \text{pro-}\mathcal{M}$ is represented by any diagram $I \to \mathcal{M}$ obtained by composing the canonical functor $\mu\backslash X \to \mathcal{M}$ with a left cofinal functor $I \to \mu\backslash X$ where I is a small left filtering; and the map $X \to UX \in \text{pro-}\mathcal{C}$ is also represented in the obvious way.

Using the above machinery, one can construct many different Artin-Mazur completions in homotopy theory; however, for our interpretation of the R-completion we shall need:

8.3 A mod-R Artin-Mazur completion in homotopy theory. Let $\mathcal{V}_0$ be the connected pointed homotopy category (i.e. the full subcategory of connected spaces in the usual (Ch.VIII, §4) pointed homotopy category); and let

$$\mathcal{M}_R \overset{\mu}{\subset} \mathcal{V}_0$$

be the full subcategory of R-nilpotent spaces with only a finite number of non-trivial homotopy groups. Then, for any $X \in \mathcal{V}_0$, the system

$$\phi\colon \quad \{X\} \longrightarrow \{(R_s X)^{(s)}\}$$

(where $\{(R_s X)^{(s)}\}$ is the "Postnikov tower" of $\{R_s X\}$) can be viewed as a tower in $\mu\backslash X$; and by 6.7 this tower is left cofinal in $\mu\backslash X$ and thus, by 8.2, the inclusion

$$\text{pro-}\mu\colon \ \text{pro-}\mathcal{M}_R \longrightarrow \text{pro-}\mathcal{V}_0$$

has a left adjoint

$$U_R\colon \ \text{pro-}\mathcal{V}_0 \longrightarrow \text{pro-}\mathcal{M}_R \,.$$

Moreover, for $X \in \mathcal{V}_0$, the Artin-Mazur completion $U_R X \in \text{pro-}\mathcal{M}_R$ is represented by the tower $\{(R_s X)^{(s)}\}$.

For our interpretation of the R-completion we also need:

8.4 Collapsing of pro-homotopy types into homotopy types. Let $\mathcal{V}_0$ be, as in 8.3, the connected pointed homotopy category, and let $\underline{X} \in \text{pro-}\mathcal{V}_0$ be a pro-object which is isomorphic in $\text{pro-}\mathcal{V}_0$ to some tower over $\mathcal{V}_0$ (This is automatic if the index filtering of $\underline{X}$ has countably many maps). Then one can collapse $\underline{X}$ to a well-defined pointed homotopy type in the obvious way, i.e. one chooses a tower of fibrations $\{Y_s\}$ over $\mathcal{S}_{*C}$ such that $\{Y_s\} \approx \underline{X} \in \text{pro-}\mathcal{V}_0$ and takes the pointed homotopy type of $\varprojlim Y_s$, which is well-defined by a version of 3.1. Unfortunately, this collapsing does not seem to be functorial in $\underline{X}$, unless one imposes stringent finiteness conditions à la [Sullivan, Ch.3].

We now conclude with the promised

8.5 Interpretation of the R-completion of a space. Combining 8.3 and 8.4, it is clear that for $X \in \mathcal{S}_{*C}$, the homotopy type of $R_\infty X$ can be obtained by:

(i) taking the Artin-Mazur completion $U_R X$, which is a "pro-homotopy type" represented by $\{(R_s X)^{(s)}\}$; and then

(ii) "collapsing" this pro-homotopy type to a homotopy type, which is represented by $\varprojlim (R_s X)^{(s)} = R_\infty X$.

An obvious defect in this approach to the R-completion is the lack of functoriality. One way around this difficulty is to impose stringent conditions, such as the finiteness conditions of [Sullivan,

Ch.3]. Another way is, <u>not</u> to work in the homotopy category; this approach is taken in Chapter XI, where we show that "<u>collapsing</u>" then becomes "<u>taking homotopy inverse limits</u>".

Chapter IV. An R-completion of groups and its relation to the R-completion of spaces

§1. Introduction

In this chapter we introduce, for any solid (Ch.I, 4.5) ring R, an Artin-Mazur-like R-completion of groups and show that it can be used to construct, up to homotopy, the R-completion of spaces. The theoretical basis for this is in Chapter III, where we developed a flexible "tower lemma" approach to R-completions. In more detail:

We define the R-completion of a group as the inverse limit of its R-nilpotent (Ch.III, 5.1) target groups. For finitely generated groups and $R = Z_p$, this R-completion reduces to the p-profinite completion of Serre, and for nilpotent groups and $R = Q$, it coincides with the Malcev completion. Like any functor on groups, this R-completion functor $\hat{-}_R$ on groups induces a functor on reduced spaces (i.e. spaces with only one vertex) as follows:

(i) Replace each reduced space X by its so-called loop group GX. This is a simplicial group, which is free in every dimension, and which has the homotopy type of "the loops on X".

(ii) Next apply the "R-completion of groups" dimension-wise to GX. This yields a simplicial group $(GX)\hat{}_R$.

(iii) Take the classifying space $\overline{W}(GX)\hat{}_R$ of the simplicial group $(GX)\hat{}_R$.

Our main result then states this classifying space $\overline{W}(GX)\hat{}_R$ has the same homotopy type as $R_\infty X$, the R-completion of X.

The chapter is organized as follows:

§2 Here we define the R-completion of a group, give various

examples, and show that the R-completion of a group B can also be obtained as the inverse limit of a tower of R-nilpotent groups which can be described in terms of the functors R_s of Ch.I, §4, namely

$$\hat{B}_R \approx \lim_{\leftarrow} \pi_1 R_s K(B,1).$$

§3 contains a variation on the R-nilpotent tower lemma of Ch.III, 6.4, which we need to efficiently formulate our main result (in §4).

§4, §6 and §7 In §4 we state the main result mentioned above. In fact, we make the slightly stronger statements that (in the sense of Ch.III, §3)

(i) for general R, the towers of fibrations

$$\{\bar{W}T_s GX\} \qquad \text{and} \qquad \{R_s X\}$$

where $T_s = \pi_1 R_s K(-,1)$, have the same weak pro-homotopy type

(ii) if $R = Z$, then the towers of fibrations

$$\{\bar{W}(GX/\Gamma_s GX)\} \qquad \text{and} \qquad \{R_s X\}$$

have the same weak pro-homotopy type (Γ_s denote the lower central series functors), and

(iii) if $R = Z_p$, then the towers of fibrations

$$\{\bar{W}(GX/\Gamma_s^{(p)} GX)\} \qquad \text{and} \qquad \{R_s X\}$$

have the same weak pro-homotopy type ($\Gamma_s^{(p)}$ denote the p-lower central series functors).

A proof of (i) which uses (ii) is given in §4, (iii) is proven in §6, and a proof of (ii) which uses (iii) is given in §7.

§5 contains some applications:

(i) A slight strengthening of the relative connectivity lemma (Ch.I, 6.2(ii)) to: "the homotopy type of $R_\infty X$ in dimensions $\leq k$" depends only on "the homotopy type of X in dimensions $\leq k$".

(ii) A first quadrant spectral sequence which, for every simplicial group B, goes to $\pi_* R_\infty \bar{W} B$.

(iii) If F is a free group, then

$$R_\infty K(F, 1) \simeq K(\hat{F}_R, 1).$$

(iv) A countable wedge of circles is Z-bad and Z_p-bad (in the sense of Ch.I, §5).

(v) A generalization of the Curtis convergence theorem to nilpotent spaces.

(vi) A generalization to fibre-wise completions.

Notation. In this chapter we will mostly work in the category $\mathcal{J}_0$ of reduced spaces, i.e. spaces with only one vertex. The reason for this is that the functors G and $\bar{W}$ are adjoint if one restricts oneself to reduced spaces, but not if one uses pointed connected spaces (in which case the functors G and $\bar{W}$ are only "adjoint up to homotopy").

Of course (Ch.I, 4.5) we again assume throughout that the ring R is solid.

§2. The R-completion of a group

We define an Artin-Mazur-like R-completion of groups, which, for finitely generated groups and $R = Z_p$ (the integers modulo a prime p), is the p-profinite completion of [Serre], and which, for nilpotent groups and $R = Q$ (the rationals) is the Malcev completion [Quillen].

It turns out that this R-completion of groups can also be described in terms of the functors R_s which we used in Ch.I, §4 to obtain the R-completion of a space.

We start with recalling from Ch.III, 5.1 the notion of

2.1 R-nilpotent groups. A group N is called R-nilpotent if N has a central series

$$N = N_1 \supset \cdots \supset N_j \supset \cdots \supset N_k = *$$

such that, for each j, the quotient group N_j/N_{j+1} admits an R-module structure. For instance:

(i) Z-nilpotent groups are groups which are nilpotent in the usual sense.

(ii) If $f\colon N \to N'$ is a group homomorphism between R-nilpotent abelian groups, then ker f and coker f are also R-nilpotent (Ch.III, 5.8).

(iii) If R_s is as in Ch.I, §4, then, for every space $X \in \mathcal{S}_{*c}$, the groups $\pi_i R_s X$ are R-nilpotent for all i and $s < \infty$ (Ch.III, 5.6).

Now we define

2.2 The R-completion of a group. The *R-completion* of a group B is the group $\hat{B}_R$ obtained by combining [Artin-Mazur, §3] with an inverse limit, i.e. by taking the inverse limit [Artin-Mazur, p.147] of the functor which assigns to every homomorphism $B \to N$, where N is R-nilpotent, the group N, and to every commutative triangle

$$\begin{array}{ccc} & & N \\ & \nearrow & \downarrow \\ B & & \\ & \searrow & \downarrow \\ & & N' \end{array}$$

with N and N' both R-nilpotent, the map $N \to N'$. The required inverse limit exists, because the above large diagram of R-nilpotent groups has a cofinal small diagram given, for example, by the tower $\{\pi_1 R_s K(B, 1)\}$ (This will follow from the proof of 2.4).

Clearly this *R-completion is a functor* and there are natural maps

$$B \xrightarrow{\phi} \hat{B}_R \qquad \text{and} \qquad \hat{\hat{B}}_{RR} \xrightarrow{\psi} \hat{B}_R$$

such that $\{\hat{-}_R, \phi, \psi\}$ *is a triple on the category of groups.*

2.3 Examples. It is not hard to see that the above definition implies:

(i) If B is R-nilpotent, then (of course)

$$\hat{B}_R \approx B.$$

(ii) If $R = Z$ (the integers), then

$$\hat{B}_R \approx \varprojlim B/\Gamma_i B$$

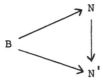

where $\{\Gamma_i B\}$ is the lower central series of B (see [Curtis (H)]). More generally one has:

(iii) Always

$$\hat{B_R} \approx \lim_{\leftarrow} (B/\Gamma_i B)\hat{_R}.$$

This follows from the fact that, for every R-nilpotent group N, there is a natural isomorphism

$$\text{Hom}\ (B, N) \approx \lim_{\rightarrow} \text{Hom}\ (B/\Gamma_i B, N)$$

(iv) If $R = Z_p$, then

$$\hat{B_R} \approx \lim_{\leftarrow} B/\Gamma_i^{(p)} B$$

where $\{\Gamma_i^{(p)} B\}$ is the p-lower central series of B (see [Rector (AS)]). A special case of this is:

(v) If $R = Z_p$ and B is finitely generated, then

$$\hat{B_R} = \text{the p-profinite completion of } B$$

of [Serre, p.I-5], and thus, if B is also abelian, then

$$\hat{B_R} \approx (\text{the p-adic integers}) \otimes B$$

(vi) If $R = Q$ and B is nilpotent, then (see Ch.V, §2)

$$\hat{B_R} = \text{the Malcev completion of } B$$

of [Malcev] and [Quillen (RH), p.279] and the map $\phi: B \to \hat{B_R}$ is universal for maps of B into nilpotent uniquely divisible groups, and thus, *if B is also abelian, then*

$$\hat{B}_R \approx R \otimes B.$$

We end with the observation that the R-completion of an arbitrary group can also be obtained as the inverse limit of a tower of R-nilpotent groups which is somewhat different than 2.3(iii) and which can be described in terms of the functors R_s of Ch.I, §4:

2.4 A reduction lemma. *Let B be a group. Then there is a natural isomorphism*

$$\hat{B}_R \approx \varprojlim \pi_1 R_s K(B, 1)$$

such that the following diagram commutes

$$\begin{array}{ccc} B & \approx & \pi_1 K(B, 1) \\ \downarrow \phi & & \downarrow \phi \\ \hat{B}_R & \approx & \varprojlim \pi_1 R_s K(B, 1). \end{array}$$

Proof. As (2.1(ii)) $\pi_1 R_s K(B, 1)$ is R-nilpotent for all $s < \infty$ it suffices to show that, for every R-nilpotent group N,

$$\varinjlim \mathrm{Hom}_{(\text{groups})}(\pi_1 R_s K(B, 1), N) \approx \mathrm{Hom}_{(\text{groups})}(B, N)$$

or equivalently that there is a 1-1 correspondence of pointed homotopy classes

$$\varinjlim [R_s K(B, 1), K(N, 1)] \approx [K(B, 1), K(N, 1)].$$

But this is the case, in view of Ch.III, 6.5 and 6.7.

§3. A variation on the R-nilpotent tower lemma

In order to be able to efficiently state the main results of this chapter (in §4) we formulate here a group-functor version of the R-nilpotent tower lemma (Ch.III, 6.4).

First we recall from [May, p.118, 87 and 122] a few facts about

3.1 The loop group functor G and the classifying functor $\bar{W}$. The loop group functor G assigns to every space $X \in \mathcal{S}_0$ (the category of reduced spaces, i.e. spaces with only one vertex) a simplicial group GX which has the homotopy type of "the loops on X", and which is a free group in each dimension. Conversely, the classifying functor $\bar{W}$ assigns to every simplicial group L a reduced space $\bar{W}L$ such that L has the homotopy type of "the loops on $\bar{W}L$". Furthermore the functors G and $\bar{W}$ are adjoint and the resulting natural map

$$X \longrightarrow \bar{W}GX \qquad \in \mathcal{S}_0$$

is a weak (homotopy) equivalence for all $X \in \mathcal{S}_0$.

To simplify the formulation of our tower lemma we next introduce the notion of

3.2 R-towers of group-functors. By this we mean a tower $\{T_s\}$ of functors and natural transformations between them

$$\cdots \longrightarrow T_s \longrightarrow T_{s-1} \longrightarrow \cdots \longrightarrow T_0 \longrightarrow T_{-1} = *$$

where each T_s is a functor

$$T_s\colon \ (\underline{\text{groups}}) \longrightarrow (\underline{\text{groups}}),$$

together with a tower map

$$\{\mathrm{Id}\} \longrightarrow \{T_s\}$$

such that for every free group F

(i) the homomorphism $T_sF \to T_{s-1}F$ $(s \geq 0)$ is onto, and its kernel is an R-module contained in the center of T_sF, and

(ii) the induced map of group homology

$$\{H_i(F;\ R)\} \longrightarrow \{H_i(T_sF;\ R)\}$$

is a pro-isomorphism (Ch.III, §2) for all $i \geq 0$.

Now we can finally state:

3.3 The group functor version of the R-nilpotent tower lemma. *Let* $\{\mathrm{Id}\} \to \{T_s\}$ *be an R-tower of group functors. Then, for every* $X \in \mathcal{J}_0$,

(i) the spaces $\overline{W}T_sGX$ *are R-nilpotent (Ch.III, 5.2), and*

(ii) the induced tower map

$$\{X\} \longrightarrow \{\overline{W}T_sGX\}$$

is an R-tower for X *(Ch.III, 6.1).*

Thus (Ch.III, 6.4) the towers of fibrations $\{\overline{W}T_sGX\}$ *and* $\{R_sX\}$ *have the same "weak pro-homotopy type" and hence*

$$\overline{W}\ \underleftarrow{\lim}\ T_sGX \ = \ \underleftarrow{\lim}\ \overline{W}T_sGX \qquad \underline{\text{and}} \qquad R_\infty X$$

<u>have the same homotopy type</u>.

<u>Proof</u>. It is not hard to see that the map $\overline{W}T_sGX \to \overline{W}T_{s-1}GX$ is a principal fibration with as fibre the simplicial R-module $\overline{W}K_sGX$, where K_sGX denotes the kernel of $T_sGX \to T_{s-1}GX$. Hence (Ch.III, 5.2 and 5.5) the spaces $\overline{W}T_sGX$ are R-nilpotent.

In order to show that the maps

$$\{H_i(X; R)\} \longrightarrow \{H_i(\overline{W}T_sGX; R)\}$$

are pro-isomorphisms we recall from [Quillen (PG), 2.1] and [Quillen (SS)] that, for a simplicial group L one can form the <u>simplicial space</u> (i.e. double simplicial set) K(L, 1) and its <u>diagonal</u> diag K(L, 1), and that

<u>(i) there is a natural first quadrant homology spectral sequence</u>

$$E^1_{p,q} = H_q(K(L_p, 1); R) \Longrightarrow H_{p+q}(\text{diag } K(L, 1); R),$$

<u>(ii) diag K(L, 1) is homotopically equivalent to $\overline{W}L$ and hence there is a natural isomorphism</u>

$$H_*(\text{diag } K(L, 1); R) \approx H_*(\overline{W}L; R).$$

One now easily proves the desired pro-isomorphism, using the map of pro-spectral sequences induced by the map $\{GX\} \to \{T_sGX\}$.

§4. A relation between the "R-completion of a group" and the "R-completion of a space"

Our main result in this chapter is, that the "R-completion of a group" can be used to obtain, at least up to homotopy, the "R-completion of a space". More precisely, any functor

$$T:\quad \underline{\text{(groups)}} \longrightarrow \underline{\text{(groups)}}$$

gives rise to a functor on reduced spaces (3.1)

$$\mathcal{S}_0 \xrightarrow{G} \underline{\text{(simplicial groups)}} \xrightarrow{T(\text{dimensionwise})} \underline{\text{(simplicial groups)}} \xrightarrow{\overline{W}} \mathcal{S}_0$$

and if T is the R-completion functor for groups (2.2), then the resulting functor $\overline{W}TG$ is homotopically equivalent to the R-completion functor for spaces R_∞. In fact, the following somewhat strong result holds:

4.1 Proposition. The map

$$\phi:\quad \{Id\} \longrightarrow \{\pi_1 R_s K(\ ,\ 1)\}$$

is an R-tower of group functors and hence (3.3), for every $X \in \mathcal{S}_0$, the towers of fibrations

$$\{\overline{W}\pi_1 R_s K(GX,\ 1)\} \qquad \underline{\text{and}} \qquad \{R_s X\}$$

have the same "weak pro-homotopy type" and (2.4) the spaces

$$\overline{W}(GX)\hat{}_R \;=\; \varprojlim \overline{W}\pi_1 R_s K(G,\ 1) \qquad \underline{\text{and}} \qquad R_\infty X$$

have the same homotopy type.

For $R = Z$ and $R = Z_p$ one also has:

4.2 Proposition. *Let* $\{\Gamma_s\}$ *denote the lower central series functors [Curtis (H)] and let* $R = Z$ *(the integers). Then the natural map*

$$\{Id\} \longrightarrow \{Id/\Gamma_s\}$$

is an R-tower of group functors and hence (3.3), for every $X \in \mathcal{S}_0$, *the towers of fibrations*

$$\{\bar{W}(GX/\Gamma_s GX)\} \quad \text{and} \quad \{R_s X\}$$

have the same "weak pro-homotopy type".

4.3 Proposition. *Let* p *be a prime, let* $\{\Gamma_s^{(p)}\}$ *denote the p-lower central series functors [Rector (AS)] and let* $R = Z_p$ *(the integers modulo p). Then the natural map*

$$\{Id\} \longrightarrow \{Id/\Gamma_s^{(p)}\}$$

is an R-tower of group functors and hence (3.3), for every $X \in \mathcal{S}_0$, *the towers of fibrations*

$$\{\bar{W}(GX/\Gamma_s^{(p)} GX)\} \quad \text{and} \quad \{R_s X\}$$

have the same "weak pro-homotopy type".

We give here a proof of proposition 4.1 which uses proposition 4.2. A proof of proposition 4.2 will be given in §7 and uses, in turn, proposition 4.3, which we prove in §6.

Proof of 4.1 (using 4.2). It is not hard to see, using Ch.II, 2.5 and Ch.III, 5.5, 5.6 and 5.8, that condition 3.2(i) is satisfied.

To prove condition 3.2(ii) we observe that, by 4.2 and Ch.III, 6.6, the map

$$\{\pi_i R_s K(F, 1)\} \longrightarrow \{\pi_i R_s K(F/\Gamma_s F, 1)\}$$

is a pro-isomorphism for all i and the following two lemmas now imply the desired result.

4.4 Lemma. If B is a torsion-free nilpotent group, then $\{\pi_i R_s K(B, 1)\}$ is pro-trivial for all $i > 1$.

Proof. The case where B is torsion-free abelian follows from the fact [Bousfield-Kan (HS), §15] that in the homotopy spectral sequence of $K(B, 1)$ (Ch.I, 4.4)

$$E_2^{s,t}(K(B, 1); R) = 0 \qquad \text{for } t-s \neq 1.$$

The general case follows from Ch.III, 3.6 because the upper central series of B has torsion free abelian quotients [Lazard, p.160].

4.5 Lemma. If $\{\pi_i R_s K(B, 1)\}$ is pro-trivial for $i > 1$, then the maps

$$\{H_i(B; R)\} \longrightarrow \{H_i(\pi_1 R_s K(B, 1); R)\}$$

are pro-isomorphisms for $i \geq 0$.

Proof. As the Postnikov map $\{R_s K(B, 1)\} \to \{(R_s K(B, 1))^{(1)}\}$ is a weak pro-homotopy equivalence, Ch.III, 3.4 implies that the maps

$$\{H_i(R_sK(B, 1); R)\} \longrightarrow \{H_i((R_sK(B, 1))^{(1)}; R)\}$$

are pro-isomorphisms and the desired result now follows from the fact that, by Ch.III, 6.5, the maps

$$\{H_i(K(B, 1); R)\} \longrightarrow \{H_i(R_sK(B, 1); R)\}$$

are also pro-isomorphisms.

§5. Applications

In this section we give several applications of 4.1.

We start with a slight strengthening of the relative connectivity lemma of Ch.I, 6.2 and show that "*the homotopy type of* $R_\infty X$ *in dimensions* $\leq k$" *depends only on "the homotopy type of* X *in dimensions* $\leq k$".

5.1 Proposition. *Let* $k \geq 0$ *and let* $f\colon X \to Y \in \mathcal{S}_0$ *be such that* $\pi_i f\colon \pi_i X \to \pi_i Y$ *is an isomorphism for* $i \leq k$ *and is onto for* $i = k+1$. *Then the induced map* $\pi_i R_\infty f\colon \pi_i R_\infty X \to \pi_i R_\infty Y$ *is also an isomorphism for* $i \leq k$ *and onto for* $i = k+1$.

Proof. We may assume that f is 1-1 in dimensions $\leq k$ and is onto in dimension $k+1$. Then $\overline{W}(Gf)\hat{}_R$ has the same properties and the desired result thus follows from 4.1.

For a simplicial group L one can apply the spectral sequence of a double simplicial group of [Quillen (SS)] to the *double simplicial group* $GK(L, 1)\hat{}_R$ and get

5.2 Proposition. *For any simplicial group* L *there is a first quadrant spectral sequence with*

$$E^2_{p,q} = \pi_q \pi_p R_\infty K(L, 1)$$

which converges to $\pi_{p+q} R_\infty \overline{W}B$.

Another immediate consequence of 4.1 is:

5.3 Proposition. Let F be a free group. Then

$$\pi_1 R_\infty K(F, 1) \approx \hat{F}_R$$

$$\pi_i R_\infty K(F, 1) = * \qquad \text{for } i \neq 1.$$

Using this we can now prove (see Ch.I, §5):

5.4 Proposition. Let F be a free group on a countable number of generators, and let $R = Z$ or $R = Z_p$. Then $K(F, 1)$ is R-bad, i.e. (Ch.I, §5) the map $H_*(K(F, 1); R) \to H_*(R_\infty K(F, 1); R)$ is not an isomorphism.

Proof. We assume here $R = Z$, but a similar proof, using the lower p-central series, works for $R = Z_p$. Writing $\hat{F}$ instead of $\hat{F}_Z$, it clearly suffices to show that the map $F/\Gamma_2 F \to \hat{F}/\Gamma_2 \hat{F}$ is not onto.

Let $x_{i,j}$ $(i \geq j)$ denote the generators of F, let

$$b = [x_{2,1}, x_{2,2}] \cdots [x_{n,1}, \cdots, x_{n,n}] \cdots \qquad \varepsilon \hat{F}$$

where $[\ ,\cdots,\ ]$ is the simple commutator, and assume that $b \in \Gamma_2 \hat{F}$, i.e. b can be written

$$b = [u_1, u_2] \cdots [u_{2k-1}, u_{2k}] \qquad u_i \in \hat{F}.$$

Let F_n denote the free group on the generators $x_{n,1}, \cdots, x_{n,n}$ and let $p_n\colon F \to F_n$ be the projection. Then

$$\hat{p}_n b = [x_{n,1}, \cdots, x_{n,n}] = [\hat{p}_n u_1, \hat{p}_n u_2] \cdots [\hat{p}_n u_{2k-1}, \hat{p}_n u_{2k}] \quad \varepsilon \hat{F}_n.$$

A straightforward computation shows that $[x_{n,1}, \cdots, x_{n,n}]$ does not lie in the subgroup of F_n generated by $\Gamma_2\Gamma_2F_n$ and $\Gamma_{n+1}F_n$ and hence there is, for each n, at least one i such that $p_n\hat{u}_i$ has non-zero image in F_n/Γ_2F_n. But this contradicts the fact that, for each j, there is only a finite number of n's such that $p_n\hat{u}_i$ has non-zero image in F_n/Γ_2F_n. Hence $b \notin \Gamma_2F^{\wedge}$.

5.5 Remark. Proposition 5.4 can certainly be improved, but we do not yet know the best possible result. It should not be hard to show that $K(F, 1)$ is R-bad for any (solid) ring R and any infinitely generated free group F. However, it is an open question whether $K(F, 1)$ is Z_p-bad or Z_p-good if F is a finitely generated free group, although we know that $K(F, 1)$ is Z-bad for some finitely generated free group F. This follows because the projective plane P^2 is Z-bad (Ch.VII, §5) and because one can show, for any (solid) ring R, that if $K(F, 1)$ were R-good for every finitely generated free group F, then all spaces $X \in \mathscr{S}$ of finite type (i.e. X_n finite for all n) would also be R-good.

Another application of §4 is the following generalization to nilpotent spaces, of

5.6 The Curtis convergence theorem. For $X \in \mathscr{S}_{*c}$ [Curtis (H), p.393] defined the lower central series spectral sequence $\{E^r_{s,t}X\}$ using the homotopy exact couple of the lower central series filtration

$$GX = \Gamma_1 GX \supset \Gamma_2 GX \supset \cdots$$

and showed that the initial term

$$E^1_{s,t}X = \pi_t(\Gamma_s GX/\Gamma_{s+1}GX) \approx \pi_t L^s(GX/\Gamma_2 GX)$$

depends only on $H_*(X;\, Z)$. Moreover [Curtis (H)] and later [Quillen (PG)] showed that, <u>for X simply connected</u>, $\{E^r_{s,t}X\}$ <u>converges strongly to</u> π_*X, thereby giving a "generalized Hurewicz theorem".

Now let $X \in \mathcal{S}_{*c}$ be <u>nilpotent</u>. Then

$$\{X\} \longrightarrow \{X\} \qquad \text{and} \qquad \{X\} \longrightarrow \{\overline{W}(GX/\Gamma_s GX)\}$$

both satisfy the conditions of the Z-nilpotent tower lemma (Ch.III, 6.4) and hence one immediately gets the following generalization of the Curtis convergence theorem to nilpotent spaces:

<u>If</u> $X \in \mathcal{S}_{*c}$ <u>is nilpotent, then the tower map</u>

$$\{\pi_t GX\} \longrightarrow \{\pi_t(GX/\Gamma_s GX)\}$$

<u>is a pro-isomorphism for</u> $t \geq 0$ <u>and hence</u> $\{E^r_{s,t}X\}$ <u>converges strongly to</u> π_*X in the following sense:

(i) For each (s,t) there exists a number $r_0(s,t)$ such that $E^r_{s,t} = E^\infty_{s,t}$ for $r > r_0(s,t)$.

(ii) For each t there exists a number $s_0(t)$ such that $E^\infty_{s,t} = 0$ for $s > s_0(t)$.

(iii) For each t, the terms $E^\infty_{1,t}X, \cdots, E^\infty_{s_0(t),t}X$ are the quotients of a finite filtration of $\pi_t GX \approx \pi_{t+1}X$.

We end with indicating how the result of 4.1 can be generalized to

<u>5.6 Fibre-wise R-completions</u>. For this one generalizes the "R-completion of a group" to a "<u>fibre-wise R-completion of a group homomorphism</u>" as follows:

For a group homomorphism $L \to M$, its <u>fibre-wise R-completion</u> will be the map $\dot{\hat{L}}_R \to M$ in the commutative diagram of groups and homomorphisms

$$\begin{array}{ccccccc} * & \longrightarrow & K & \longrightarrow & L & \longrightarrow & M \\ & & \downarrow{\scriptstyle\phi} & & \downarrow & & \downarrow{\scriptstyle =} \\ * & \longrightarrow & K^{\wedge}_R & \longrightarrow & \dot{L}^{\wedge}_R & \longrightarrow & M \end{array}$$

in which both rows are exact and in which $\dot{L}^{\wedge}_R$ is the group with generators

$$(l,h) \qquad\qquad l \in L,\ h \in K^{\wedge}_R$$

and relations

$$(lk,h) = (l,kh) \qquad\qquad k \in K,\ l \in L,\ h \in K^{\wedge}_R$$

$$(l,h)(l',h') = (ll',l'(h)h') \qquad\qquad l,l' \in L,\ h,h' \in K^{\wedge}_R$$

where, for $k \in K$, we denote by the same symbol its images in L and $K^{\wedge}_R$, and where, for $l' \in L$, we use the same symbol to denote the automorphism of $K^{\wedge}_R$, which is the R-completion of the automorphism of K which, in turn, is the restriction of the inner automorphism of L induced by l'.

Using this it is now not hard to construct, for a fibration $X \to B \in \mathcal{J}_0$ with fibre F, a fibration with B as base, $\overline{W}(GF)^{\wedge}_R$ as fibre and of which the total space

$$\overline{W}(G\dot{X})^{\wedge}_R \quad \times_{\overline{W}GB} \quad B$$

has the same homotopy type as the fibre-wise R-completion $\dot{R}_\infty X$ of Ch.I, §8.

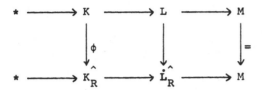

§6. Proof of proposition 4.3

We have to prove that *for every free group* F *the map*

$$\{H_i(F;\ Z_p)\} \longrightarrow \{H_i(F/\Gamma_s^{(p)}F;\ Z_p)\}$$

is a pro-isomorphism for all $i \geq 0$.

For this we need the following result of [Stallings]:

6.1 Lemma. *If* N *is a normal subgroup of a group* B, *then there is a natural exact sequence*

$$H_2(B;\ Z_p) \rightarrow H_2(B/N;\ Z_p) \rightarrow N/(B\#N) \rightarrow H_1(B;\ Z_p) \rightarrow H_1(B/N;\ Z_p) \rightarrow *$$

where $B\#N$ *is the subgroup of* N *generated by all elements of the form*

$$bnb^{-1}n^{-1} \qquad b \in B,\ n \in N$$

$$n^p \qquad n \in N.$$

Using this we prove:

6.2 Lemma. *If* F *is a free group, then the map*

$$H_2(F/\Gamma_{ps}^{(p)}F;\ Z_p) \longrightarrow H_2(F/\Gamma_s^{(p)}F;\ Z_p)$$

is zero for all $s \geq 1$.

Proof. Applying 6.1 to $\Gamma_s^{(p)}F \subset F$ one obtains

$$H_2(F/\Gamma_s^{(p)}F;\ Z_p) \approx \Gamma_s^{(p)}F/(F\#\Gamma_s^{(p)}F)$$

and the lemma now follows from the fact that $\Gamma_{ps}^{(p)}F \subset F\#\Gamma_s^{(p)}F$.

Next we show

6.3 Lemma. If F is a finitely generated free group, then the map

$$\{H_i(F;\ Z_p)\} \longrightarrow \{H_i(F/\Gamma_s^{(p)}F;\ Z_p)\}$$

is a pro-isomorphism for all $i \geq 0$.

Proof. Since

$$H_i(F;\ Z_p) \approx H_i(F/\Gamma_s^{(p)}F;\ Z_p) \qquad i = 0, 1$$

$$= * \qquad i \geq 2$$

it suffices to prove for $i \geq 2$

$(6.3)_i$: *The pro-abelian group* $\{H_i(F/\Gamma_s^{(p)}F;\ M_s)\}$ *is pro-trivial for any tower* $\{M_s\}$ *such that*

(a) each M_s *is a module over the* Z_p*-group ring of* $F/\Gamma_s^{(p)}F$,

(b) each M_s *is finitely generated as a* Z_p*-module*,

(c) each tower map $M_s \to M_{s-1}$ *is equivariant with respect to* $F/\Gamma_s^{(p)}F \to F/\Gamma_{s-1}^{(p)}F$.

To verify that $(6.3)_i$ implies $(6.3)_{i+1}$ for $i \geq 2$, let R_s denote the Z_p-group ring of the finite group $F/\Gamma_s^{(p)}F$ and define

K_s by the short exact sequence

$$0 \longrightarrow K_s \longrightarrow R_s \otimes M_s \xrightarrow{\text{mult.}} M_s \longrightarrow 0$$

of R_s-modules. Then

$$\{H_{i+1}(F/\Gamma_s^{(p)}F;\ M_s)\} \approx \{H_i(F/\Gamma_s^{(p)}F;\ K_s)\}$$

since $H_n(F/\Gamma_s^{(p)}F;\ R_s \otimes M_s) = 0$ for $n > 0$. It thus suffices to show $(6.3)_2$.

For this let $I_s \subset R_s$ be the augmentation ideal and take the filtration

$$M_s \supset I_sM_s \supset \cdots \supset (I_s)^jM_s \supset \cdots .$$

Since the finite p-group $F/\Gamma_s^{(p)}F$ must act nilpotently on the finite abelian p-group M_s [Hall, p.47], it follows that $(I_s)^qM_s = 0$ for some q, depending on s. Moreover 6.2 implies that

$$\{H_2(F/\Gamma_s^{(p)}F;\ (I_s)^jM_s/(I_s)^{j+1}M_s)\}$$

is pro-trivial for each j, since the coefficients are not twisted and $(6.3)_2$ now readily follows.

To generalize lemma 6.3 to proposition 4.3 we need the notion of

<u>6.4 Functors of finite degree</u>. A functor

$$T\colon \ \underline{\text{(pointed sets)}} \longrightarrow \underline{\text{(abelian groups)}}$$

is said to be *of finite degree* if

(i) $T^* = 0$, and

(ii) there is an integer $d \geq 0$ such that, for every pointed set X, the abelian group TX is generated by the subgroups $TX_\alpha \subset TX$, where $X_\alpha \subset X$ runs through the pointed subsets of X with $\leq d$ elements other than $*$.

This definition readily implies:

6.5 Proposition. If T is of finite degree and $a \in TX$ is non-zero. Then there is a map of pointed sets $f: X \to Y$ such that

(i) $(Tf)a \neq 0$, and

(ii) Y is finite.

Proposition 4.3 now follows easily from this proposition, 6.3 and

6.6 Lemma. Let

$$F: \text{(pointed sets)} \longrightarrow \text{(groups)}$$

denote the functor which assigns to a pointed set X the free group generated by X with the relation $* = 1$. Then the functors

$$H_i(FX/\Gamma_s^{(p)}FX;\ Z_p)$$

are of finite degree for all i and s.

Proof. This is verified by induction on s, using the Hochschild-Serre homology spectral sequence for

$$\Gamma_s^{(p)}FX/\Gamma_{s+1}^{(p)}FX \longrightarrow FX/\Gamma_{s+1}^{(p)}FX \longrightarrow FX/\Gamma_s^{(p)}FX$$

and the fact that each of the functors

$$H_i(\Gamma_s^{(p)}FX/\Gamma_{s+1}^{(p)}FX;\ Z_p)$$

is of finite degree because $\Gamma_s^{(p)}FX/\Gamma_{s+1}^{(p)}FX \approx L_s^{(p)}(Z_pX)$ where $L_s^{(p)}(Z_pX)$ is the s-th component of the free restricted Lie algebra (see [Rector (AS)]) generated by Z_pX.

§7. Proof of proposition 4.2 (using 4.3)

We have to prove that, for every free group F, the maps

$$\{H_i(F; Z)\} \longrightarrow \{H_i(F/\Gamma_s F; Z)\}$$

are pro-isomorphisms for all $i \geq 0$.

To do this recall from [Curtis (L)] that

$$\Gamma_s F/\Gamma_{s+1}F \approx L_s(F/\Gamma_2 F) \qquad s \geq 1$$

where L_s is the s-th component of the free Lie ring functor. The argument of the second half of §6 then shows that one only has to consider finitely generated F. In that case, however, each of the groups $H_i(F/\Gamma_s F; Z)$ is also finitely generated, and thus it suffices to show that the maps

$$\{H_i(F; R)\} \longrightarrow \{H_i(F/\Gamma_s F; R)\} \qquad i \geq 0$$

are pro-isomorphisms for $R = Q$ and $R = Z_p$, p prime, or equivalently (Ch.III, 6.6):

7.1 Lemma. If F is a free group, then the maps

$$\{\pi_i R_s K(F, 1)\} \longrightarrow \{\pi_i R_s K(F/\Gamma_s F, 1)\} \qquad i \geq 1$$

are pro-isomorphisms for $R = Q$ and $R = Z_p$, p prime.

This is an easy consequence of the following four lemmas:

7.2 Lemma. For every group B and (solid) ring R, the map

$$\{\pi_1 R_s K(B, 1)\} \longrightarrow \{\pi_1 R_s K(B/\Gamma_s B, 1)\}$$

is a pro-isomorphism.

Proof. As (2.1(iii)) the groups $\pi_1 R_s K(B, 1)$ and $\pi_1 R_s K(B/\Gamma_s B, 1)$ are nilpotent for $0 \le s < \infty$, it suffices to show that, for every R-nilpotent group N,

$$\varinjlim \mathrm{Hom}_{(\mathrm{groups})}(\pi_1 R_s K(B, 1), N) \approx \mathrm{Hom}_{(\mathrm{groups})}(B, N)$$

$$\varinjlim \mathrm{Hom}_{(\mathrm{groups})}(\pi_1 R_s K(B/\Gamma_s B, 1), N) \approx \mathrm{Hom}_{(\mathrm{groups})}(B, N)$$

and this is an easy consequence of 2.4 and the easily established fact that

$$\varinjlim \mathrm{Hom}_{(\mathrm{groups})}(B/\Gamma_s B, N) \approx \mathrm{Hom}_{(\mathrm{groups})}(B, N).$$

7.3 Lemma. If F is a free group and R is a (solid) ring, then $\{\pi_i R_s K(F/\Gamma_s F, 1)\}$ is pro-trivial for $i > 1$.

This follows immediately from 4.4:

7.4 Lemma. If F is a free group and $R = Z_p$ (p prime), then $\{\pi_i R_s K(F; 1)\}$ is pro-trivial for $i > 1$.

Proof. By 4.3 the maps

$$\{H_i(F; R)\} \longrightarrow \{H_i(F/\Gamma_s^{(p)} F; R)\} \qquad i \ge 1$$

are pro-isomorphisms and hence, by Ch.III, 6.6, so are the maps

$$\{\pi_i R_s K(F, 1)\} \longrightarrow \{\pi_i R_s K(F/\Gamma_s^{(p)} F, 1)\} \qquad i \geq 1.$$

Since $K(F/\Gamma_s^{(p)}, 1)$ is R-nilpotent, it follows from Ch.III, 5.3 that each

$$\{\pi_i R_t K(F/\Gamma_s^{(p)} F, 1)\} \qquad i, s > 1$$

is pro-trivial and this readily implies the lemma.

7.5 Lemma. *If F is a free group and $R = Q$, then $\{\pi_i R_s K(F, 1)\}$ is pro-trivial for $i > 1$.*

Proof. It was shown in [Bousfield-Kan (PP), 15.6] that for a pointed connected space X the homotopy spectral sequence (Ch.I, 4.4) satisfies

$$E_2(X;\, Q) \approx P\, \mathrm{Cotor}^{H_*(X;\, Q)}(Q, Q)$$

where P denotes the primitive element functor. It follows that

$$E_2^{s,t}(K(F, 1);\, Q) \approx L_{s+1} H_1(K(F, 1);\, Q) \qquad t-s = 1$$

$$= 0 \qquad t-s \neq 1$$

where L_{s+1} is as in the beginning of this section. This easily implies the lemma.

Chapter V. R-localizations of nilpotent spaces

§1. Introduction

The main purpose of this chapter is to show that, for $R \subset Q$, the R-completion of the preceding chapters is a localization with respect to a set of primes, and that therefore various well-known results about localizations of simply connected spaces remain valid for nilpotent spaces (i.e. spaces for which, up to homotopy, the Postnikov tower can be refined to a tower of principal fibrations).

In more detail:

§2 contains some algebraic preliminaries and deals with a Malcev completion $R \otimes N$, which is defined for every nilpotent group N and ring $R \subset Q$, and which we use quite frequently in this chapter.

§3 Here we prove that, for a nilpotent space X, the homotopy (and integral homology) groups of the R-completion of X are the Malcev completions of the homotopy (and integral homology) groups of X. We also list some easy consequences of this and obtain a strong convergence result for the homotopy spectral sequence $\{E_r(X; R)\}$ (Ch.I, 4.4) of a nilpotent space X.

§4 In this section we observe that the main result of §3 implies that the usual notion of R-localization for simply connected spaces generalizes to nilpotent spaces and, moreover, is merely an "up to homotopy" version of the R-completion. We also recall several basic properties of localizations.

§5, §6 and §7 The last two of these sections deal with a prime fracture lemma and a fracture square lemma, which both state that, under suitable conditions, a homotopy classification problem can be split into a "rational problem" and "problems involving various primes or sets of primes". In preparation for the proof of these fracture lemmas, we generalize (in §5) the main result of §3 to function spaces.

§8 Here we use the prime fracture lemma to obtain a prime fracturing of H-space structures for H-spaces, which have the integral homology of a finite space (but need not be simply connected).

§9 discusses the Zabrodsky mixing of nilpotent spaces and how this can be used, in conjunction with the fracture square lemma, to construct H-spaces which have the integral homology of a finite space.

Notation. In this chapter we will mainly work in the category $\mathcal{S}_{*C}$ of pointed connected spaces and in its full subcategory $\mathcal{S}_{*N}$ of pointed connected nilpotent spaces.

And, of course, throughout this chapter, even when we forget to mention it, the ring R will always be a subring of the rationals Q.

§2. Malcev completions for nilpotent groups

We discuss here, for every nilpotent group N and ring $R \subset Q$, a Malcev completion $R \otimes N$, which will be used in subsequent sections to describe homotopy groups and other invariants of R-completions of nilpotent spaces. This terminology is justified by the fact that, for $R = Q$, the group $R \otimes N$ is the Malcev completion of [Malcev] and [Quillen (RH)]; we use the notation $R \otimes N$ because the functor $R\otimes -$ behaves very much like an ordinary tensor product and, in fact, for abelian N, is an ordinary tensor product.

The algebraic results of this section are well-known for $R = Q$ [Malcev], [Quillen (RH)]; for $R \neq Q$ they are implicit in [Lazard]. However, with our present machinery, we can avoid the complicated algebra used by these authors.

We start with two propositions:

2.1 Proposition. *If N is an abelian group and $R \subset Q$, then there are natural isomorphisms*

$$\pi_1 R_\infty K(N, 1) \approx \hat{N}_R \approx R \otimes N.$$

Moreover $\{\pi_i R_s K(N, 1)\}$ is pro-trivial for all $i > 1$ and hence

$$\pi_i R_\infty K(N, 1) = * \qquad \text{for } i \neq 1.$$

Proof. One proves, as in Ch.IV, 4.4, that $\{\pi_i R_s K(N, 1)\}$ is pro-trivial for $i > 1$. The rest of the proposition then readily follows from Ch.IV, 2.2 and 2.4.

Combining this with Ch.IV, 2.4 and Ch.III, 3.6 one gets:

2.2 Proposition. *If N is a nilpotent group and $R \subset Q$, then $\hat{N}_R$ is R-nilpotent (Ch.III, 5.1) and there is a natural isomorphism*

$$\pi_1 R_\infty K(N, 1) \approx \hat{N}_R .$$

Moreover $\{\pi_i R_s K(N, 1)\}$ is pro-trivial for $i > 1$ and hence

$$\pi_i R_\infty K(N, 1) = * \qquad \text{for } i \neq 1.$$

In view of this we now define

2.3 The Malcev completion of a nilpotent group (for $R \subset Q$). For a nilpotent group N and ring $R \subset Q$ we define the Malcev completion $R \otimes N$ by

$$R \otimes N = \hat{N}_R$$

or equivalently (2.2)

$$R \otimes N = \pi_1 R_\infty K(N, 1).$$

Clearly $R \otimes N$ is natural in R and N and comes with a natural map

$$\phi\colon N \longrightarrow R \otimes N$$

induced by the natural map $\phi\colon N \to \hat{N}_R$ or equivalently the natural map $\phi\colon K(N, 1) \to R_\infty K(N, 1)$.

The rest of this section is devoted to showing that $R \otimes N$ has all the expected properties:

2.4 Exactness property. Every short exact sequence of nilpotent groups

$$* \longrightarrow N' \longrightarrow N \longrightarrow N'' \longrightarrow *$$

induces a short exact sequence

$$* \longrightarrow R \otimes N' \longrightarrow R \otimes N \longrightarrow R \otimes N'' \longrightarrow * .$$

Proof. This follows from 2.6, since, by Ch.III, 4.4 and 4.8

$$R_\infty K(N', 1) \longrightarrow R_\infty K(N, 1) \longrightarrow R_\infty K(N'', 1)$$

is, up to homotopy, a fibration.

2.5 Universal property.

(i) $R \otimes N$ is R-nilpotent (2.2),

(ii) the map $\phi: N \to R \otimes N$ is universal for maps from N to R-nilpotent groups, and

(iii) the map $\phi: N \to R \otimes N$ is an isomorphism if and only if N is R-nilpotent; in particular, the map $N \to Z \otimes N$ is always an isomorphism.

The proof is straightforward.

To give this universal property the same form as that of [Quillen (RH), p.278] for $R = Q$, we define:

<u>2.6 Uniquely J-divisible groups</u>. Let J be a set of primes. A group G then is called <u>uniquely J-divisible</u>, if, for each prime $p \in J$, the p-th power function

$$-^p: G \longrightarrow G$$

is a bijection. Then one has the following

<u>2.7 Characterization of R-nilpotent groups (for $R \subset Q$). Let J be a set of primes, and let $R = Z[J^{-1}]$ (Ch.I, §9). A nilpotent group N then is uniquely J-divisible if and only if it is R-nilpotent.</u>

<u>Proof</u>. It suffices to prove:

(i) <u>If, in a central group extension</u>

$$* \longrightarrow G' \longrightarrow G \longrightarrow G'' \longrightarrow *$$

<u>two of the groups are uniquely J-divisible, then so is the third.</u>

(ii) <u>The center of a nilpotent uniquely J-divisible group is also uniquely J-divisible.</u>

Statement (i) is trivial and (ii) follows from the observation of [Lazard, p.159] that $[x^r, y] = 1$ <u>implies</u> $[x, y]^{r^{(c-1)}} = 1$, where x and y are elements in a nilpotent group of class $\leq c$ and [,] denotes the commutator.

<u>2.8 The kernel and image of the map $\phi: N \to R \otimes N$</u>. <u>Let J be a set of primes and let $R = Z[J^{-1}]$ (Ch.I, §9). Then</u>

<u>(i) the kernel of $\phi: N \to R \otimes N$ consists of the J-torsion in N</u>, i.e. the elements $u \in N$ such that $u^r = 1$ for some integer r

of which all the prime factors are in J, and

(ii) for every $v \in R \otimes N$, there exists an integer $r \geq 1$, *of which all the prime factors are in J, such that v^r is in the image of the map $\phi\colon N \to R \otimes N$.*

Proof. This follows by an easy inductive argument, using the ladders of central extensions

$$\begin{array}{ccccccccc}
* & \longrightarrow & \Gamma_s N/\Gamma_{s+1} N & \longrightarrow & N/\Gamma_{s+1} N & \longrightarrow & N/\Gamma_s N & \longrightarrow & * \\
 & & \downarrow & & \downarrow & & \downarrow & & \\
* & \longrightarrow & R \otimes \Gamma_s N/\Gamma_{s+1} N & \longrightarrow & R \otimes N/\Gamma_{s+1} N & \longrightarrow & R \otimes N/\Gamma_s N & \longrightarrow & *
\end{array}$$

where the Γ_s denote the lower central series functors (see [Curtis]).

2.9 Proposition. If $R, R' \subset Q$, then the obvious map (2.5)

$$R \otimes (R' \otimes N) \longrightarrow (R \otimes R') \otimes N$$

is an isomorphism.

Proof. This is obvious if N is abelian and the general case follows inductively.

§3. Homotopy and homology properties of the R-completion of a nilpotent space $(R \subset Q)$

Our main purpose is to show:

3.1 Proposition. *If* $X \in \mathcal{S}_{*N}$ (i.e. X is pointed connected and Z-nilpotent in the sense of Ch.III, §5) *and, of course,* $R \subset Q$, *then*

(i) $R_\infty X$ *is R-nilpotent and so are the groups*

$$\pi_* R_\infty X \qquad \text{and} \qquad \tilde{H}_*(R_\infty X; Z)$$

(ii) The canonical maps (2.5)

$$R \otimes \pi_* X \longrightarrow \pi_* R_\infty X \qquad \text{and} \qquad R \otimes \tilde{H}_*(X; Z) \longrightarrow \tilde{H}_*(R_\infty X; Z)$$

are isomorphisms.

(Note that an abelian group is R-nilpotent if and only if it is an R-module).

Before proving this we list some easy consequences for the homotopy and homology properties of $R_\infty X$. It turns out that the roles of π_* and $\tilde{H}_*(-; Z)$ are symmetric. We also obtain a strong convergence result for the homotopy spectral sequence $\{E_r(X; R)\}$ (Ch.I, 4.4) of a nilpotent space.

3.2 Proposition. *For a map* $f: X \to Y \in \mathcal{S}_{*N}$, *the following statements are equivalent:*

(i) f *induces a homotopy equivalence* $R_\infty X \simeq R_\infty Y$.

(ii) f *induces an isomorphism* $R \otimes \pi_* X \approx R \otimes \pi_* Y$.

(iii) f induces an isomorphism $\tilde{H}_*(X; R) \approx \tilde{H}_*(Y; R)$.

(Note that $\tilde{H}_*(X; R) \approx R \otimes \tilde{H}_*(X; Z)$).

Proof. In view of Ch.I, 5.5, (i) is equivalent to (iii) and, in view of 3.1, (i) is equivalent to (ii).

3.3 Proposition. *For a space* $X \in \mathcal{S}_{*N}$, *the following are equivalent*:

(i) X is R-complete (i.e. $X \simeq R_\infty X$).

(ii) The groups $\pi_n X$ *are R-nilpotent*.

(iii) The groups $\tilde{H}_n(X; Z)$ *are R-nilpotent*.

(iv) Whenever a map $f: K \to L \in \mathcal{S}_*$ *induces an isomorphism* $H_*(K; R) \approx H_*(L; R)$, *then it induces a bijection* $[L, X] \approx [K, X]$ *of pointed homotopy classes of maps (Ch.VIII, §4)*.

Proof. This is an easy consequence of 3.1 and 3.2, and Ch.II, 2.8.

3.4 Proposition. *Every space* $X \in \mathcal{S}_{*N}$ *is R-good, i.e. the natural map* $\tilde{H}_*(X; R) \to \tilde{H}_*(R_\infty X; R)$ *is an isomorphism*.

This is immediate from 3.1.

Combining 3.4 with Ch.II, 2.8 one gets:

3.5 Proposition. *If* $X \in \mathcal{S}_{*N}$, *then the natural map* $\phi: X \to R_\infty X$ *induces, for every* $W \in \mathcal{S}_{*C}$, *an isomorphism of pointed homotopy classes of maps (Ch.VIII, §4)*

$$[R_\infty X, R_\infty W] \approx [X, R_\infty W].$$

3.6 Example. For $R \subset Q$ and $n \geq 1$ there is an isomorphism

$$[R_\infty S^n, R_\infty S^n] \approx \pi_n R_\infty S^n \approx R$$

which assigns to each map $f: R_\infty S^n \to R_\infty S^n$ the obvious *degree* $\deg f \in R$.

We conclude with a

3.7 Curtis convergence theorem for $R \subset Q$. *If* $X \in \mathcal{S}_{*N}$ *and* $R \subset Q$, *then the homotopy spectral sequence* $\{E_r(X; R)\}$ *(Ch.I, 4.4) converges strongly to* $\pi_* R_\infty X \approx R \otimes \pi_* X$ *in the following sense:*

(i) $\{E_r(X; R)\}$ *is Mittag-Leffler (Ch.IX, 5.5) in all dimensions* ≥ 1.

(ii) For each $i \geq 1$, *there exists a number* $s_0(i)$ *such that*

$$E_\infty^{s,s+i}(X; R) = 0 \qquad \textit{for} \quad s > s_0(i).$$

(iii) For each $i \geq 1$, *the terms*

$$E_\infty^{0,i}(X; R), \cdots, E_\infty^{s_0(i),\, s_0(i)+i}(X; R)$$

are the quotients of a finite filtration of

$$\pi_i R_\infty X \approx R \otimes \pi_i X.$$

This convergence result was initially proved in [Bousfield-Kan (HS), §6] for simply connected spaces by combining Curtis' fundamental theorem (Ch.IV, 5.6 and [Curtis (H)]) with some ad-hoc simplicial arguments. Our present approach is much more direct; it is essentially the same as our generalization to nilpotent spaces (Ch.IV, 5.6) of Curtis' original theorem.

Proof of 3.7. If $R = Z$, then (Ch.III, 6.4) the map $\{X\} \to \{R_s X\}$ is a weak pro-homotopy equivalence and this immediately implies the desired result.

The general case now follows from the fact that $E_r(X; R) \approx R \otimes E_r(X; Z)$ for $R \subset Q$ and $1 \leq r < \infty$.

Finally we give the

Proof of 3.1. First consider the case that $X = K(G, n)$ with G abelian. Then the obvious map $K(G, n) \to K(R \otimes G, n)$ induces an isomorphism on R-homology and the desired result follows from Ch.I, 5.5 and Ch.II, 2.7.

Next suppose that $F \to E \to B \in \mathcal{S}_{*C}$ is a principal fibration and that the proposition is already proved for $X = F$ and $X = B$. Then $R_\infty F \to R_\infty E \to R_\infty B \in \mathcal{S}_{*C}$ is (by Ch.II, 2.2), up to homotopy, a principal fibration and thus (Ch.III, 5.5) $R_\infty E$ is R-nilpotent. It follows that $\pi_* R_\infty E$ is R-nilpotent and, using the obvious homotopy exact sequences, one shows that $R \otimes \pi_* E \approx \pi_* R_\infty E$. Using the obvious Serre spectral sequences one shows that $\tilde{H}_*(E; R) \approx \tilde{H}_*(R_\infty E; R)$ and that $\tilde{H}_*(R_\infty E; Z_p) = 0$ for each $p \in J$, where $R = Z[J^{-1}]$. Hence $\tilde{H}_*(R_\infty E; Z)$ is R-nilpotent and $R \otimes \tilde{H}_*(E; Z) \approx \tilde{H}_*(R_\infty E; Z)$. Thus the proposition holds for $X = E$.

The general case now follows from Ch.I, 6.2.

§4. R-localizations of nilpotent spaces $(R \subset Q)$

Following [Quillen (RH)], [Sullivan], [Mimura-Nishida-Toda] and others in the simply connected case, we introduce the notion of an R-localization of a nilpotent space, and show that it is merely an "up to homotopy" version of our R-completion. We then deduce that the R-completion preserves, up to homotopy, various basic constructions and end this section with an easy example of an R-localization, obtained as an infinite mapping cylinder.

4.1 R-localizations. For $X \in \mathcal{S}_{*N}$ and, of course, $R \subset Q$, an R-localization of X is a map $X \to \overline{X} \in \mathcal{S}_{*N}$ such that either of the following (equivalent by §3) conditions hold:

(i) The groups $\pi_*\overline{X}$ are R-nilpotent and the canonical map $R \otimes \pi_*X \to \pi_*\overline{X}$ is an isomorphism.

(ii) The groups $\tilde{H}_*(\overline{X}; Z)$ are R-nilpotent and the canonical map $R \otimes \tilde{H}_*(X; Z) \to \tilde{H}_*(\overline{X}; Z)$ is an isomorphism.

The results of §3 then immediately imply:

4.2 Proposition. R-localization is well-defined and functorial on the pointed homotopy category of nilpotent spaces. It is induced by the functor R_∞.

4.3 Homotopy characterization of $R_\infty X$. For $X \in \mathcal{S}_{*N}$ (and, of course, $R \subset Q$), the R-completion $X \to R_\infty X$ is an R-localization, and, in the pointed homotopy category, any R-localization $X \to \overline{X}$ is canonically equivalent to $X \to R_\infty X$.

Next we show that, up to homotopy, the R-completion preserves various constructions $(R \subset Q)$.

We already have, from Ch.II, 4.4' and 4.7 a

4.4 Fibre lemma. *Let* $p\colon E \to B \in \mathcal{S}_{*N}$ *be a fibration with* $\pi_1 E \to \pi_1 B$ *onto. Then* $R_\infty p\colon R_\infty E \to R_\infty B$ *is a fibration, and the inclusion* $R_\infty(p^{-1}*) \to (R_\infty p)^{-1}*$ *is a homotopy equivalence. Moreover the fibre* $p^{-1}* \in \mathcal{S}_{*N}$.

There is a corresponding

4.5 Cofibre lemma. *Let* $i\colon A \to X \in \mathcal{S}_{*N}$ *be a cofibration (i.e. injection) with* $\pi_1 A \to \pi_1 X$ *onto. Then* $R_\infty i\colon R_\infty A \to R_\infty X$ *is a cofibration and the obvious map* $R_\infty X/R_\infty A \to R_\infty(X/A)$ *is a weak equivalence.*

Proof. It is easy to check that R_∞ always preserves cofibrations. Since $\pi_1 A \to \pi_1 X$ and $R \otimes \pi_1 A \to R \otimes \pi_1 X$ are onto, it follows that the cofibres X/A and $R_\infty X/R_\infty A$ are 1-connected. A homology argument now shows that $X/A \to R_\infty X/R_\infty A$ is an R-localization, so the lemma follows from 4.3.

Some more "preservation properties" of R_∞ $(R \subset Q)$ are given in

4.6 Proposition. *If* $X, Y \in \mathcal{S}_{*N}$ *then, in the pointed homotopy category, there are canonical equivalences*

(i) $R_\infty(SX) \simeq SR_\infty X$, *where* S *denotes the suspension [May, p.124]*

(ii) $R_\infty(\Omega X) \simeq \Omega R_\infty X$, *where* Ω *denotes the loop functor [May, p.99] and* X *is 1-connected.*

(iii) $R_\infty(X \times Y) \simeq R_\infty X \times R_\infty Y$

(iv) $R_\infty(X \vee Y) \approx R_\infty X \vee R_\infty Y$, where X and Y are 1-connected.

(v) $R_\infty(X \wedge Y) \approx R_\infty X \wedge R_\infty Y$

Proof. The obvious maps

$$SX \longrightarrow SR_\infty X \qquad\qquad X \times Y \longrightarrow R_\infty X \times R_\infty Y$$

$$X \vee Y \longrightarrow R_\infty X \vee R_\infty Y$$

$$\Omega X \longrightarrow \Omega R_\infty X \qquad\qquad X \wedge Y \longrightarrow R_\infty X \wedge R_\infty Y$$

are clearly R-localizations, so the proposition follows from 4.3.

R-localizations can often be constructed by direct limit methods using:

4.7 Infinite mapping cylinders. For an infinite sequence of maps

$$X^0 \xrightarrow{f^0} X^1 \longrightarrow \cdots \longrightarrow X^n \xrightarrow{f^n} \cdots \qquad \in \mathcal{J}_{*C}$$

the infinite mapping cylinder is the space $X^\infty \in \mathcal{J}_{*C}$ obtained from the disjoint union of the pointed mapping cylinders $\coprod_n M(f^n)$, by identifying for all n

$$X^n \subset M(f^n) \qquad \text{with} \qquad X^n \subset M(f^{n-1})$$

It is easy to show that the inclusions $X^n \subset X^\infty$ induce natural isomorphisms

$$\varinjlim \pi_* X^n \approx \pi_* X^\infty$$

$$\varinjlim \tilde{H}_*(X^n; G) \approx \tilde{H}_*(X^\infty; G) \qquad \text{for } G \text{ abelian.}$$

We end with applying this to an example of an

4.8 R-localization for loop spaces and suspensions. Let $X \in \mathcal{S}_{*N}$ be fibrant (i.e. $X \to *$ is a fibration) and assume that X has the homotopy type of a loop space (resp. a suspension). Then, for each positive integer n, there is a map $n\colon X \to X \in \mathcal{S}_{*N}$, which induces "multiplication by n"

$$n\colon \pi_* X \longrightarrow \pi_* X \qquad (\text{resp. } n\colon \tilde{H}_*(X; Z) \longrightarrow \tilde{H}_*(X; Z)).$$

Now let $n_1, n_2, n_3, \cdots$ be a sequence of positive integers such that the prime factors of each n_i lie in J (where $R = Z[J^{-1}]$), and each prime in J occurs as a factor of infinitely many n_i. Then it is easy to show that the inclusion of $X^0 = X$ in the infinite mapping cylinder X^∞ of

$$X^0 = X \xrightarrow{n_1} X \xrightarrow{n_2} X \xrightarrow{n_3} \cdots$$

is an R-localization, i.e. $X^\infty \simeq R_\infty X$.

This construction, of course, works also for H-spaces and nilpotent co-H-spaces.

§5. R-localizations of function spaces

In preparation for the fracture lemmas (§6) we show here that the homotopy types of the pointed function spaces (Ch.VIII, §4)

$$\hom_*(W, X) \qquad \text{and} \qquad \hom_*(W, R_\infty X)$$

are often closely related.

We start with a proposition which implies that, under suitable conditions, the R-completion of any component of $\hom_*(W, X)$ has the same homotopy type as the corresponding component of $\hom_*(W, R_\infty X)$.

5.1 Proposition. *Let $X \in \mathcal{S}_{*N}$ be fibrant, let $W \in \mathcal{S}_{*C}$ be finite (i.e. have a finite number of non-degenerate simplices) and let, of course, $R \subset Q$. Then, for every map $f: W \to X \in \mathcal{S}_{*C}$ and all $i \geq 1$,*

(i) $\pi_i(\hom_*(W, X), f)$ is nilpotent,

(ii) $\pi_i(\hom_*(W, R_\infty X), \phi f)$ is R-nilpotent, and

(iii) the map $\phi: X \to R_\infty X$ induces an isomorphism

$$R \otimes \pi_i(\hom_*(W, X), f) \approx \pi_i(\hom_*(W, R_\infty X), \phi f)$$

To prove this we need

5.2 Lemma. *If $E \to B \in \mathcal{S}$ is a fibration such that every component of E is nilpotent, then every component of every fibre is also nilpotent.*

Proof. Choose a base point $* \in E$ and let F be the fibre containing it. Then $\pi_1 F$ acts on the resulting long exact homotopy sequence and the desired result follows easily from Ch.II, 4.2.

Proof of 5.1. We may assume that W is *reduced* (i.e. has only one vertex) and show first, by induction on the skeletons of W, that *every component of* $\hom_*(W, X)$ *is nilpotent*. Clearly

$$\hom_*(W^{[0]}, X) = *.$$

Furthermore the map

$$\hom_*(W^{[k]}, X) \longrightarrow \hom_*(W^{[k-1]}, X), \qquad k \geq 1,$$

is, up to homotopy, a fibration induced from the obvious map

$$\hom_*(W^{[k-1]}, X) \longrightarrow \hom_*(V, X)$$

where V is a wedge of boundaries of standard k-simplices (Ch.VIII, 2.12), and the desired result follows from lemma 5.2.

The rest of the proposition is now easy to prove, using 3.1, 2.4 and, again, induction on the skeletons of W.

The relation between the sets of components of $\hom_*(W, X)$ and $\hom_*(W, R_\infty X)$, i.e. the relation between the *pointed homotopy classes of maps* (Ch.VIII, §4)

$$[W, X] \qquad \text{and} \qquad [W, R_\infty X]$$

is not so easy to describe. Of course one has

5.3 Proposition. *Let* $X \in \mathcal{S}_{*N}$ *be fibrant, let* $W \in \mathcal{S}_{*C}$ *be finite and let either* W *be a reduced suspension [May, p.124] or* X *be a homotopy associative H-space (Ch.I, 7.5). Then*

(i) [W, X] is a nilpotent group,

(ii) $[W, R_\infty X]$ is an R-nilpotent group, and

(iii) the map $\phi: X \to R_\infty X$ induces an isomorphism

$$R \otimes [W, X] \approx [W, R_\infty X].$$

Proof. If W is a reduced suspension, then the proof goes as in 5.1.

If X is a homotopy associative H-space, then [Stasheff (H), p.10] X has a homotopy inverse and thus [G.W. Whitehead] [W, X] is a nilpotent group. Furthermore, by Ch.I, 7.5, $R_\infty X$ is also a homotopy associative H-space. The rest of the proof proceeds as in 5.1.

In general, however, the sets [W, X] and $[W, R_\infty X]$ do not come with a group structure. Still it is possible to make some useful statements (5.5) by observing that for every map $f: W \to X$ there are subsets

$$[W, X]_f \subset [W, X] \qquad \text{and} \qquad [W, R_\infty X]_{\phi f} \subset [W, R_\infty X]$$

which have a group structure and which we call

5.4 Neighborhood groups. Let $X \in \mathcal{S}_{*c}$ be fibrant, let $W \in \mathcal{S}_{*c}$ be finite and reduced and let $n = \dim W$ (i.e. W has at least one non-degenerate n-simplex and $W^{[n]} = W$). Then one has, up to homotopy, a fibration

$$\hom_*(W, X) \xrightarrow{j} \hom_*(W^{[n-1]}, X) \xrightarrow{p} \hom_*(V, X)$$

where V is a wedge of boundaries of standard n-simplices (Ch.VIII, 2.12) and thus, for every map $f: W \to X$, the corresponding long exact homotopy sequence

$$\cdots \longrightarrow \pi_1(\hom_*(W^{[n-1]}, X), f|W^{[n-1]}) \xrightarrow{p_*} \pi_1(\hom_*(V, X), f|V) \xrightarrow{\partial}$$

$$\xrightarrow{\partial} \pi_0(\hom_*(W, X), f) \xrightarrow{j_*} \pi_0(\hom_*(W^{[n-1]}, X), f|W^{[n-1]}) \longrightarrow .$$

Using this we now define the neighborhood group $[W, X]_f$ of f to be the group

$$[W, X]_f = \operatorname{coker} p_*$$

which is abelian if $n \geq 2$; and which, as a set, is also given by

$$[W, X]_f = \ker j_* \subset \pi_0 \hom_*(W, X) = [W, X]$$

i.e. $[W, X]_f$ consists of all $u \in [W, X]$ such that

$$u|W^{[n-1]} \simeq f|W^{[n-1]}.$$

Using 5.1 and the definition of $[W, X]_f$ it is now not hard to prove:

5.5 Proposition. *Let $X \in \mathcal{S}_{*N}$ be fibrant and let $W \in \mathcal{S}_{*C}$ be finite and reduced. Then, for every map $f: W \to X \in \mathcal{S}_{*C}$,*

(i) $[W, R_\infty X]_{\phi f}$ is an R-nilpotent group, and

(ii) the map $\phi: X \to R_\infty X$ induces an isomorphism

$$R \otimes [W, X]_f \approx [W, R_\infty X]_{\phi f}.$$

5.6 Remark. It has long been recognized that Brown's representability theorem can be used to define localizations for certain H-spaces. Although we will not pursue this idea, we note that proposition 5.3 implies that the R-localization of an Ω-spectrum

<u>corresponds to the R-tensoring of the associated cohomology theory for finite spaces</u>.

§6. Fracture lemmas

"Fracture lemmas" show that, under suitable conditions, a homotopy classification problem can be split into a "rational problem" and "problems involving various primes or sets of primes". They yield many of the interesting applications of localizations.

The first satisfactory fracture lemma seems to have been proved by Sullivan in the context of his completion theory [Sullivan, Ch.3], and our approach was inspired by his work. Also [Hilton-Mislin-Roitberg] have independently proved fracture lemmas by methods somewhat similar to ours.

6.1 Notation. For a set I of primes, let $Z_{(I)}$ denote the *integers localized at* I, i.e. (Ch.I, 9.3)

$$Z_{(I)} = Z[J^{-1}]$$

where J consists of all primes not in I; and, for $X \in \mathcal{S}$, let

$$X_{(I)} = (Z_{(I)})_\infty X \qquad \in \mathcal{S}.$$

In particular

$$Z_{(0)} = Q \qquad\qquad X_{(0)} = Q_\infty X.$$

Then we have the

6.2 Prime fracture lemma. *Let* $X \in \mathcal{S}_{*N}$ *have finitely generated homotopy groups, let* $W \in \mathcal{S}_{*C}$ *be finite* (i.e. have a finite number of non-degenerate simplices) *and let* I *be a set of primes.*

Then the natural map of pointed homotopy classes of maps (Ch.VIII, §4)

$$\Phi: \quad [W, X_{(I)}] \longrightarrow \underset{p \in I}{\text{pull-back}} \{[W, X_{(p)}] \longrightarrow [W, X_{(0)}]\}$$

where p *ranges over all primes in* I*, is an isomorphism.*

The most interesting case of 6.2 occurs when I consists of all primes and thus $[W, X_{(I)}] \approx [W, X]$.

6.3 Fracture square lemma. *Let* I *and* J *be sets of primes, let* $X \in \mathcal{S}_{*N}$ *and let* $W \in \mathcal{S}_{*C}$ *be finite. Then*

(i) the natural diagram

$$\begin{array}{ccc} X_{(I \cup J)} & \longrightarrow & X_{(I)} \\ \downarrow & & \downarrow \\ X_{(J)} & \longrightarrow & X_{(I \cap J)} \end{array}$$

is, up to homotopy, a fibre square, and

(ii) the natural square of pointed homotopy classes of maps (Ch.VIII, §4)

$$\begin{array}{ccc} [W, X_{(I \cup J)}] & \longrightarrow & [W, X_{(I)}] \\ \downarrow & & \downarrow \\ [W, X_{(J)}] & \longrightarrow & [W, X_{(I \cap J)}] \end{array}$$

is a pull-back.

6.4 Remark. In view of the fracture square lemma one can, for $X \in \mathcal{S}_{*N}$ and any *finite* partition $I_1, \cdots, I_n$ of the primes, recover the homotopy type of X from

(i) the homotopy types of $X_{(I_1)}, \cdots, X_{(I_n)}$, and

(ii) the *rational* information contained in the homotopy equivalences

$$X_{(I_1)(0)} \simeq \cdots \simeq X_{(I_n)(0)}.$$

One can*not* dispense with this last ingredient since, for instance, the "Hilton-Roitberg criminal" and $Sp(2)$ have homotopically equivalent localizations at the prime 2 and at the odd primes [Mislin].

Also the homotopy type of X is usually *not* recovered if one takes the pull-back of the fibrations corresponding to

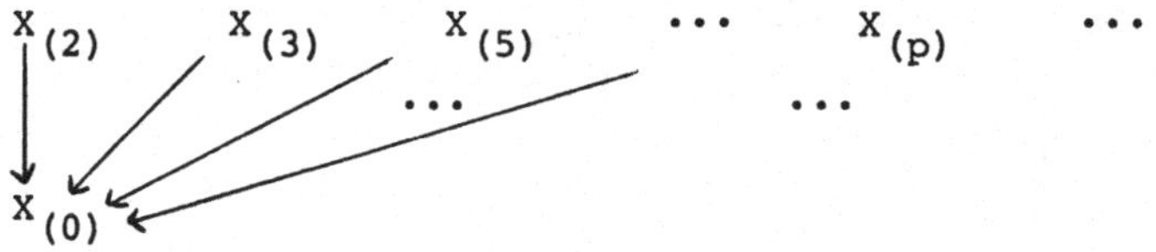

p prime

A counter example already occurs when $X = K(Z, n)$, because of a $\varprojlim^1$ term (see Ch.IX).

To prove the above fracture lemmas we need their *group theoretic analogues*:

6.5 Lemma. *If N is a finitely generated nilpotent group and I is a set of primes, then the natural map*

$$Z_{(I)} \otimes N \longrightarrow \underset{p \in I}{\text{pull-back}} \{Z_{(p)} \otimes N \longrightarrow Q \otimes N\}$$

where p ranges over all primes in I, is an isomorphism.

Proof. The lemma clearly is true when N is finitely generated abelian. Moreover, if

$$* \longrightarrow N' \longrightarrow N \longrightarrow N'' \longrightarrow *$$

is a short exact sequence of finitely generated nilpotent groups and the lemma holds for N' and N'' then, in view of 2.4, it also holds for N. This readily implies the general case.

A similar argument shows:

6.6 Lemma. *If I and J are sets of primes and N is a nilpotent group, then the natural diagram*

$$\begin{array}{ccc} Z_{(I\cup J)} \otimes N & \longrightarrow & Z_{(I)} \otimes N \\ \downarrow & & \downarrow \\ Z_{(J)} \otimes N & \longrightarrow & Z_{(I\cap J)} \otimes N \end{array}$$

is a pull-back. Moreover every element $u \in Z_{(I\cap J)} \otimes N$ *can be expressed as* $u = vw$ *where* v *(resp.* w*) is in the image of* $Z_{(I)} \otimes N$ *(resp.* $Z_{(J)} \otimes N$*).*

Proof of 6.2. We can assume that W is *reduced* (i.e. has only one vertex) and we will prove 6.2 by induction on the skeletons $W^{[n]}$ of W. Thus assuming that

$$\phi^n\colon\ [W^{[n]}, X_{(I)}] \longrightarrow \underset{p \in I}{\underline{\text{pull-back}}}\ \{[W^{[n]}, X_{(p)}] \longrightarrow [W^{[n]}, X_{(0)}]\}$$

is an isomorphism for $n = k-1$, we have to show that this is also the case for $n = k$.

To show that Φ^k <u>is injective</u>, we suppose $f, g: W^{[k]} \to X_{(I)}$ with $\Phi^k[f] = \Phi^k[g]$ and thus, by our inductive hypothesis $[f]|W^{[k-1]} = [g]|W^{[k-1]}$. Since the element $[g] \in [W^{[k]}, X_{(I)}]_f$ (5.4) goes to zero under the obvious map

$$\Psi: \quad [W^{[k]}, X_{(I)}]_f \longrightarrow \underset{p \in I}{\underline{\text{pull-back}}} \{[W^{[k]}, X_{(p)}]_{\phi f} \longrightarrow [W^{[k]}, X_{(0)}]_{\phi f}\}$$

and since, by 5.5 and 6.5, Ψ is an isomorphism, it follows that $[g] = 0 \in [W^{[k]}, X_{(I)}]_f$. Hence $[g] = [f] \in [W^{[k]}, X_{(I)}]$, and thus Φ^k is injective.

To show that Φ^k <u>is surjective</u>, we suppose

$$h \in \underset{p \in I}{\underline{\text{pull-back}}} \{[W^{[k]}, X_{(p)}] \longrightarrow [W^{[k]}, X_{(0)}]\}.$$

By our inductive assumption, there exists a map $d: W^{[k-1]} \to X_{(I)}$ such that $\Phi^{k-1}[d] = h|W^{[k-1]}$. The map d has an extension $e: W^{[k]} \to X_{(I)}$ since the obstruction to extending d lies in a finite product of copies of $\pi_k X_{(I)}$ and since (3.1)

$$\pi_k X_{(I)} \approx \underset{p \in I}{\underline{\text{pull-back}}} \{\pi_k X_{(p)} \longrightarrow \pi_k X_{(0)}\}.$$

Let $g \in [W^{[k]}, X_{(I)}]_e$ denote the element corresponding to, h under the isomorphism

$$\Psi: \quad [W^{[k]}, X_{(I)}]_e \longrightarrow \underset{p \in I}{\underline{\text{pull-back}}} \{[W^{[k]}, X_{(p)}]_{\phi e} \longrightarrow [W^{[k]}, X_{(0)}]_{\phi e}\}$$

Then it is easy to check that $g \in [W^{[k]}, X_{(I)}]$ satisfies $\Phi^k g = h$, and thus Φ^k is surjective.

<u>Proof of 6.3</u>. Part (i) is an easy consequence of 3.1 and 6.6, while part (ii) follows by a proof similar to that of 6.2. Of course, the surjectivity of the map from $[W, X_{(I \cup J)}]$ to the pull-back can also be deduced from part (i).

6.7 Remark. When $X \in \mathcal{S}_{*C}$ is a *homotopy associative H-space* (Ch.I, 7.5) or W is a *reduced suspension* [May, p.124], then one can use 5.3 (instead of 5.5) to give an easy proof of the fracture lemmas 6.2 and 6.3.

§7. Some slight generalizations

The fracture lemmas 6.2 and 6.3 are certainly not best possible: although it is not clear whether the restrictions on X can be relaxed, both lemmas obviously hold for many spaces W which are not finite. For instance, one clearly has:

7.1 Proposition. If $f: W' \to W \in \mathcal{S}_{*C}$ is such that $H_*(f; Z)$ is an isomorphism, and if the fracture lemmas 6.2 and 6.3 hold for W', then (Ch.II, 2.8) they also hold for W.

7.2 Proposition. If the fracture lemmas 6.2 and 6.3 hold for some $W \in \mathcal{S}_{*C}$, then they also hold for any space dominated by W.

7.3 Proposition. If the fracture lemmas 6.2 and 6.3 hold for a set of spaces $W_a \in \mathcal{S}_{*C}$, then they also hold for their wedge $\bigvee_a W_a$.

A useful consequence of 7.1 is the following which is also not hard to prove:

7.4 Proposition. Let $W \in \mathcal{S}_{*C}$ be of finite type (i.e. each W_k is finite) and suppose there is an integer n such that $H_i(W; Z) = 0$ for $i > n$. Then there exists a map $f: W' \to W \in \mathcal{S}_{*C}$ such that W' is finite and $H_*(f; Z)$ is an isomorphism, and hence the fracture lemmas 6.2 and 6.3 hold for W.

The usefulness of this proposition is due to the following lemma, or actually its corollary 7.6:

7.5 Lemma. For a space $W \in \mathcal{S}_{*N}$ *the following conditions are equivalent:*

(i) W *has the (weak) homotopy type of a space of finite type*

(ii) $H_i(W; Z)$ *is finitely generated for each* $i \geq 1$

(iii) $\pi_i W$ *is finitely generated for each* $i \geq 1$.

7.6 Corollary. If $W \in \mathcal{S}_{*N}$ *has the (integral) homology of a finite complex, then the fracture lemmas 6.2 and 6.3 hold for* W.

Proof of 7.5. *(i)* $\Longrightarrow$ *(ii).* This is obvious

(ii) $\Longrightarrow$ *(iii).* Since each $H_i(X; Z)$ is finitely generated, so is each of the groups

$$E^1_{s,t}W \approx \pi_t L^s(GW/\Gamma_2 GW)$$

in the lower central series spectral sequence (Ch.IV, §5). Since, for nilpotent W, this spectral sequence converges strongly to $\pi_* W$ (Ch.IV, §5), it follows that each $\pi_i W$ is finitely generated.

(iii) $\Longrightarrow$ *(i).* Since $\pi_1 W$ is finitely generated and nilpotent, it is finitely presentable [P. Hall, p.426] and its integral group ring is left and right Noetherian [P. Hall, p.429]. Since the universal cover $\tilde{W}$ is simply connected with finitely generated homotopy groups, each $H_i(\tilde{W}; Z)$ is finitely generated. Thus [Wall, p.58 and p.61] implies (i).

We end this section with observing that, while (7.3) "finite homological dimensionality" is not a necessary condition, the following counter example indicates that it is *not* enough to assume that W be merely of finite type.

7.7 Counter example. If I and J are non-empty complementary sets of primes, then the square

$$\begin{array}{ccc} [P^\infty C, S^3] & \longrightarrow & [P^\infty C, S^3_{(J)}] \\ \downarrow & & \downarrow \\ [P^\infty C, S^3_{(I)}] & \longrightarrow & [P^\infty C, S^3_{(0)}] \end{array}$$

is not a pull-back.

To see this, it suffices (Ch.IX, 3.3) to show that the obvious map

$$\varprojlim{}^1 [SP^n C, S^3] \longrightarrow \varprojlim{}^1 [SP^n C, S^3_{(I)}] \oplus \varprojlim{}^1 [SP^n C, S^3_{(J)}]$$

is not injective. Taking the $\varprojlim^*$ exact sequence (Ch.IX, 2.3) of the short exact sequence of abelian group towers

$$0 \to [SP^n C, S^3] \to [SP^n C, S^3_{(I)}] \oplus [SP^n C, S^3_{(J)}] \to [SP^n C, S^3_{(0)}] \to 0$$

it now suffices to show that $\varprojlim [SP^n C, S^3_{(0)}]$ is not generated by the images of $\varprojlim [SP^n C, S^3_{(I)}]$ and $\varprojlim [SP^n C, S^3_{(J)}]$. For this purpose, note that $SP^1 C \approx S^3$ and consider the restriction maps

$$\varprojlim [SP^n C, S^3_{(0)}] \longrightarrow [S^3, S^3_{(0)}] \approx Q$$

$$\varprojlim [SP^n C, S^3_{(I)}] \longrightarrow [S^3, S^3_{(I)}] \approx Z_{(I)}$$

$$\varprojlim [SP^n C, S^3_{(J)}] \longrightarrow [S^3, S^3_{(J)}] \approx Z_{(J)}.$$

Since $S^3_{(0)}$ represents $H^3(-; Q)$, the first map is an isomorphism, so it suffices to show that the images of the other two maps do not generate Q. This is easily proved using the non-triviality of the

Steenrod operation

$$P^1: \; H^3(SP^{\infty}C;\ Z_p) \longrightarrow H^{3+2p}(SP^{\infty}C;\ Z_p)$$

for all primes p.

§8. Fracturing H-space structures

In this section we discuss the fact that, under suitable conditions, the problem whether a space has an H-space structure, can be fractured into mod-p problems (see [Mislin]).

We start with some remarks on

8.1 H-spaces and quasi Hopf algebras. If $X \in \mathcal{J}_{*C}$ is an H-space (Ch.I, 7.5) with $H^*(X; Q)$ of finite type, then clearly

$$H^*(X; Q) \qquad (\approx H^*(X_{(I)}; Q) \quad \text{for all } I)$$

is a connected quasi Hopf algebra [Milnor-Moore, p.232]. This quasi Hopf algebra completely determines the homotopy type and the homotopy class of the H-space structure of the localization $X_{(0)}$; in fact one even has the somewhat stronger result that:

The functor $H^*(-; Q)$ is an equivalence between the category of Q-nilpotent H-spaces for which $H^*(-; Q)$ is of finite type (i.e. the category with as objects the connected Q-nilpotent H-spaces for which $H^*(-; Q)$ is of finite type, and as maps the homotopy classes of maps which are compatible with the H-space structures) and the category of connected quasi Hopf algebras over Q of finite type, which have a commutative and associative multiplication.

This is not hard to prove once one observes that the Borel theorem of [Milnor-Moore, p.255] implies that these quasi Hopf algebras are free as algebras and that therefore, as algebras, they are the cohomology of a product of $K(Q, n)$'s.

We end with the comment that clearly a Q-nilpotent H-space $X \in \mathcal{J}_{*C}$ is homotopy associative if and only the quasi Hopf algebra $H^*(X; Q)$ is a Hopf algebra.

An easy consequence of Ch.I, 7.5, the prime fracture lemma 6.2 and its generalization 7.6 now is the

8.2 Prime fracture lemma for H-spaces. *Let* $X \in \mathcal{S}_{*N}$ *have the integral homology of a finite space and let*

$$\Delta: \; H^*(X;\, Q) \longrightarrow H^*(X;\, Q) \otimes H^*(X;\, Q)$$

be a quasi Hopf algebra comultiplication. Then X *has an H-space structure inducing* Δ *if and only if, for each prime* p*, the space* $X_{(p)}$ *has an H-space structure inducing* Δ *under the canonical isomorphism* $H^*(X_{(p)};\, Q) \approx H^*(X;\, Q)$.

8.3 Remark. In order that $H^*(X;\, Q)$ has a quasi Hopf algebra comultiplication, it is necessary that $H^*(X;\, Q)$ be an exterior algebra [Milnor-Moore, p.255]. Moreover, if $H^*(X;\, Q)$ is an exterior algebra, then there is a *unique* comultiplication map making $H^*(X;\, Q)$ a Hopf algebra. Hence 8.2 has the following refinement:

8.4 Proposition. *Let* $X \in \mathcal{S}_{*N}$ *have the integral homology of a finite space. Then* X *has a homotopy associative H-space structure if and only if* $X_{(p)}$ *has a homotopy associative H-space structure for every prime* p.

We end with an

8.5 Example [Adams (S)]. Consider the n-sphere S^n for n *odd*. If p is an odd prime, then $S^n_{(p)}$ has an H-space multiplication given by

$$S^n_{(p)} \times S^n_{(p)} \xrightarrow{f \times id} S^n_{(p)} \times S^n_{(p)} \xrightarrow{g} S^n_{(p)}$$

where f is of degree $1/2$ and g is induced by a map $S^n \times S^n \to S^n$ of degree $(2,1)$ (see [Steenrod-Epstein, p.14]). Thus the problem whether S^n is an H-space is purely a mod-2 problem, and, of course, the Hopf invariant theorem shows that $S^n_{(2)}$ is an H-space if and only if $n = 1, 3, 7$. Note also that the H-space structure on $S^n_{(p)}$ has the obvious implication that Whitehead products in $\pi_* S^n$ vanish when either factor is of odd order.

§9. Zabrodsky mixing of nilpotent spaces

In recent years, localization methods have played a central role in the construction of new H-spaces (e.g. [Zabrodsky] and [Stasheff]) and the basic tool in this work has been *Zabrodsky mixing* [Zabrodsky].

9.1 Zabrodsky mixing of nilpotent spaces. Let the primes be partitioned into two disjoint sets I and J and let

$$f\colon X \longrightarrow W \qquad \text{and} \qquad g\colon Y \longrightarrow W \qquad \in \mathcal{S}_{*N}$$

be maps which induce isomorphisms

$$Q \otimes \pi_* X \approx Q \otimes \pi_* W \qquad\qquad Q \otimes \pi_* Y \approx Q \otimes \pi_* W.$$

Then, in the notation of 6.1, the *Zabrodsky mixing* $M \in \mathcal{S}_*$ *of* $X_{(I)}$ *with* $Y_{(J)}$ *over* $W_{(0)}$ is the homotopy inverse limit (i.e., Ch.XI, the "dual" to the double mapping cylinder) of the diagram

$$X_{(I)} \xrightarrow{\ f_{(\ )}\ } W_{(0)} \xleftarrow{\ g_{(\ )}\ } Y_{(J)}$$

induced by f and g. This means that, up to homotopy, there is a *pointed fibre square*

$$\begin{array}{ccc} M & \longrightarrow & Y_{(J)} \\ \downarrow & & \downarrow {\scriptstyle g_{(\ )}} \\ X_{(I)} & \xrightarrow{\ f_{(\ )}\ } & W_{(0)} \end{array}$$

The point of this construction is that "*M looks like X over the primes I*" and "*M looks like Y over the primes J*". More precisely:

9.2 Lemma. The space $M \in \mathcal{S}_*$ is connected and nilpotent, the obvious maps

$$Z_{(I)} \otimes \pi_* M \longrightarrow \pi_* X_{(I)} \qquad\qquad Z_{(J)} \otimes \pi_* M \longrightarrow \pi_* Y_{(J)}$$

are isomorphisms and hence the obvious maps

$$M_{(I)} \longrightarrow X_{(I)} \qquad\qquad M_{(J)} \longrightarrow Y_{(J)}$$

are homotopy equivalences.

Also, Zabrodsky mixings have the virtue:

9.3 Lemma. If X and Y have the integral homology of a finite space, then, assuming of course the hypotheses of 9.1, so does the Zabrodsky mixing M.

Before proving these lemmas we show how Zabrodsky mixing can be used to create new H-spaces.

9.4 Proposition. Let I and J be complementary sets of primes, let $X, Y \in \mathcal{S}_{*N}$ have the integral homology of a finite space and suppose that $X_{(I)}$ and $Y_{(J)}$ are H-spaces and that the induced quasi Hopf algebras (8.1) $H^*(X; Q)$ and $H^*(Y; Q)$ are isomorphic. Then there exists an H-space $M \in \mathcal{S}_{*N}$ which has the integral homology of a finite space and is such that, as H-spaces

$$M_{(I)} \simeq X_{(I)} \qquad \text{and} \qquad M_{(J)} \simeq Y_{(J)}.$$

Proof. This follows readily from 8.1, 9.2, 9.3 and the fracture square lemma 6.3 and its generalization 7.6.

9.5 Remarks. Of course, this proposition applies when X and Y are connected H-spaces with the integral homology of a finite space, such that $H^*(X; Q)$ and $H^*(Y; Q)$ are Hopf algebras with the same number of generators in each dimension. However, it is also useful when X and Y are not themselves H-spaces.

The above Zabrodsky mixing technique has a number of refinements and variants; for instance, one can mix classifying spaces to create new examples of finite loop spaces.

Proof of 9.2. We first claim that, for $n \geq 1$, *each element $u \in \pi_n W_{(0)}$ can be expressed as a product $u = vw$, where v and w are in the respective images of $\pi_n X_{(I)}$ and $\pi_n Y_{(J)}$.* This follows, since 2.8 shows the existence of relatively prime integers s and t such that u^s and u^t are in the respective images of $\pi_n X_{(I)}$ and $\pi_n Y_{(J)}$.

This claim implies that M is connected and that, for $n \geq 1$, the square

$$\begin{array}{ccc} \pi_n M & \longrightarrow & \pi_n Y_{(J)} \\ \downarrow & & \downarrow \\ \pi_n X_{(I)} & \longrightarrow & \pi_n W_{(0)} \end{array}$$

is *exact* in the sense that it is a pull-back such that every element of $\pi_n W_{(0)}$ factors as in the claim. Using the obvious action of $\pi_1 M$ on this exact square it is easy to show that M is nilpotent, and since (2.4) $Q \otimes -$ preserves exactness for a square of nilpotent groups, it is clear that $Q \otimes \pi_n M \approx \pi_n W_{(0)}$. And finally, since the obvious map of exact squares

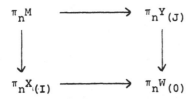

$$\begin{array}{ccc} \pi_n M & \longrightarrow & Z_{(J)} \otimes \pi_n M \\ \downarrow & & \downarrow \\ Z_{(I)} \otimes \pi_n M & \longrightarrow & Q \otimes \pi_n M \end{array} \quad \Longrightarrow \quad \begin{array}{ccc} \pi_n M & \longrightarrow & \pi_n Y_{(J)} \\ \downarrow & & \downarrow \\ \pi_n X_{(I)} & \longrightarrow & \pi_n W_{(0)} \end{array}$$

is an isomorphism on the initial and terminal corners, it is an isomorphism on all corners.

Proof of 9.3. In view of 3.1 and 9.2 each

$$Z_{(I)} \otimes H_i(M;\ Z) \qquad \textbf{resp.} \qquad Z_{(J)} \otimes H_i(M;\ Z)$$

is finitely generated as a $Z_{(I)}$-module, resp. a $Z_{(J)}$-module, and so each $H_i(M;\ Z)$ is a finitely generated abelian group. Moreover, by the same argument, for sufficiently large i

$$Z_{(I)} \otimes H_i(M;\ Z) = 0 = Z_{(J)} \otimes H_i(M;\ Z)$$

and hence $H_i(M;\ Z) = 0$.

Chapter VI. p-completions of nilpotent spaces

§1. Introduction

In this chapter we discuss the p-completion, i.e. the "up to homotopy" version of the Z_p-completion, for nilpotent spaces. It turns out that this p-completion is closely related to the p-profinite completion of [Quillen (PG)] and [Sullivan, Ch.3]; indeed, one can show that these completions coincide for spaces with Z_p-homology of finite type, although they differ for more general spaces. The basic properties of p-profinite completions are well-known for simply connected spaces of finite type, and the main purpose of this chapter is to obtain similar results for p-completions of arbitrary nilpotent spaces.

The organization of this chapter is similar to that of Chapter V.

§2, §3 and §4 contain some algebraic preliminaries. In §2 we define for every nilpotent group N and prime p, an

Ext completion $\operatorname{Ext}(Z_{p^\infty}, N)$ and a

Hom completion $\operatorname{Hom}(Z_{p^\infty}, N)$

and we show in §3 that

$\operatorname{Ext}(Z_{p^\infty}, N)$ is "really" a completion of N

$\operatorname{Hom}(Z_{p^\infty}, N)$ is nothing but $\operatorname{Hom}_{groups}(Z_{p^\infty}, N)$.

Various examples of Ext and Hom completions are discussed in §4.

§5 and §6 Our key result in §5 is that, for a nilpotent space X and $R = Z_p$, there are splittable exact sequences

$$* \longrightarrow \mathrm{Ext}(Z_{p^\infty}, \pi_n X) \longrightarrow \pi_n R_\infty X \longrightarrow \mathrm{Hom}(Z_{p^\infty}, \pi_{n-1} X) \longrightarrow *.$$

In §6 we use this result to introduce a notion of p-completion for nilpotent spaces, which is merely an "up to homotopy" version of the Z_p-completion and which generalizes the usual p-profinite completion for simply connected spaces of finite type. We also list several basic properties of p-completions.

§7 and §8 In §7 we generalize the main result of §5 to function spaces and then use this in §8 to prove an arithmetic square fracture lemma, which states that, under suitable conditions, a homotopy classification problem can be split into "Z_p-problems" and a "rational problem".

§9 contains convergence results for the homotopy spectral sequence $\{E_r(X; Z_p)\}$ of a nilpotent space X (Ch.I, 4.4).

Notation. In this chapter we again work mainly in the category $\mathcal{S}_{*C}$ of pointed connected spaces and its full subcategory $\mathcal{S}_{*N}$ of pointed connected nilpotent spaces.

And, of course, throughout this chapter, even when we forget to mention it, the ring R will always be $R = Z_p$ (p prime).

§2. Ext-completions and Hom-completions for nilpotent groups

We introduce here, for every nilpotent group N and prime p, an

Ext-completion $\quad \operatorname{Ext}(Z_{p^\infty}, N)$ and an

Hom-completion $\quad \operatorname{Hom}(Z_{p^\infty}, N)$

which will be used in subsequent sections, to describe homotopy groups and other invariants of Z_p-completions of nilpotent spaces.

For N abelian, $\operatorname{Ext}(Z_{p^\infty}, N)$ and $\operatorname{Hom}(Z_{p^\infty}, N)$ will, of course, be the usual groups, where Z_{p^∞} denotes the p-primary component of Q/Z, while for a general nilpotent group N we will have

$$\operatorname{Ext}(Z_{p^\infty}, N) = \pi_1 R_\infty K(N, 1)$$

$$\operatorname{Hom}(Z_{p^\infty}, N) = \pi_2 R_\infty K(N, 1)$$

where $R = Z_p$. In the important case of a nilpotent group N, whose p-torsion elements are of bounded order this will imply that

$$\operatorname{Ext}(Z_{p^\infty}, N) \simeq \hat{N}_{Z_p} \qquad \operatorname{Hom}(Z_{p^\infty}, N) = *$$

where $\hat{N}_{Z_p}$ is the Z_p-completion of N of Ch.IV, 2.2.

In the abelian case several algebraists [Harrison], [Rotman], [Stratton] have previously studied the "total Ext-completion"

$$\operatorname{Ext}(Q/Z, N) = \prod_p \operatorname{Ext}(Z_{p^\infty}, N).$$

We begin by reviewing the somewhat familiar

2.1 Ext and Hom completions of abelian groups from an algebraic point of view. *For abelian groups the functors*

$$\mathrm{Ext}(Z_{p^\infty}, -) \qquad \text{and} \qquad \mathrm{Hom}(Z_{p^\infty}, -)$$

are resp. the 0th *and* 1st *left derived functors of the* Z_p*-completion functor* $(\ )^\wedge_{Z_p}$ *on abelian groups of Ch.IV, 2.2.*

Proof. To prove this observe that

$$Z_{p^\infty} = \varinjlim Z_{p^n}$$

i.e. Z_{p^∞} is the direct limit of the monomorphisms

$$Z_{p^n} \longrightarrow Z_{p^{n+1}} \qquad \text{induced by} \qquad Z \xrightarrow{p} Z.$$

Then, for abelian N,

$$\varprojlim \mathrm{Ext}(Z_{p^n}, N) \approx \varprojlim N/p^nN \approx N^\wedge_{Z_p}$$

and hence [Roos, Th.1] *there is a natural short exact sequence*

$$* \longrightarrow \varprojlim{}^1 \mathrm{Hom}(Z_{p^n}, N) \longrightarrow \mathrm{Ext}(Z_{p^\infty}, N) \longrightarrow N^\wedge_{Z_p} \longrightarrow *$$

and thus, by Ch.IX, 2.2, *if* N *is an abelian group, whose p-torsion elements are of bounded order, then*

$$\mathrm{Ext}(Z_{p^\infty}, N) \approx N^\wedge_{Z_p} \qquad\qquad \mathrm{Hom}(Z_{p^\infty}, N) = *$$

and the desired result now follows from the fact that *a short exact sequence of abelian groups*

$$* \longrightarrow N' \longrightarrow N \longrightarrow N'' \longrightarrow *$$

gives rise to a natural exact sequence

$$* \longrightarrow \mathrm{Hom}(Z_{p^\infty}, N') \longrightarrow \mathrm{Hom}(Z_{p^\infty}, N) \longrightarrow \mathrm{Hom}(Z_{p^\infty}, N'') \longrightarrow$$

$$\longrightarrow \mathrm{Ext}(Z_{p^\infty}, N') \longrightarrow \mathrm{Ext}(Z_{p^\infty}, N) \longrightarrow \mathrm{Ext}(Z_{p^\infty}, N'') \longrightarrow * .$$

Next we look at the

2.2 Ext and Hom completions of abelian groups from a homotopical point of view. *For an abelian group* N *and* $R = Z_p$, *there are natural isomorphisms*

$$\pi_1 R_\infty K(N, 1) \approx \mathrm{Ext}(Z_{p^\infty}, N)$$

$$\pi_2 R_\infty K(N, 1) \approx \mathrm{Hom}(Z_{p^\infty}, N)$$

$$\pi_i R_\infty K(N, 1) = * \qquad \textit{for } i \neq 1, 2$$

such that the following diagrams commute:

(i) the diagram

$$\begin{array}{ccc} \pi_1 R_\infty K(N, 1) & \approx & \mathrm{Ext}(Z_{p^\infty}, N) \\ \downarrow & & \downarrow \\ \varprojlim \pi_1 R_s K(N, 1) & \approx & N^\wedge_{Z_p} \end{array}$$

where the map on the left is as in Ch.IX, §3 and the bottom isomorphism is as in Ch.IV, 2.4,

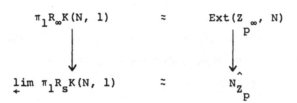

<u>(ii) for every short exact sequence of abelian groups $* \to N' \to N \to N'' \to *$, the diagram</u>

$$\begin{array}{ccc} \pi_2 R_\infty K(N'', 1) & \approx & \operatorname{Hom}(Z_{p^\infty}, N'') \\ \downarrow & & \downarrow \\ \pi_1 R_\infty K(N', 1) & \approx & \operatorname{Ext}(Z_{p^\infty}, N') \end{array}$$

<u>where the map on the left is defined using Ch.II, 2.2.</u>

<u>Proof</u>. If N is free abelian, then, by Ch.IX, 3.1 and Ch.IV, 2.4 and 4.4

$$\begin{aligned} \pi_i R_\infty K(N, 1) &\approx \hat{N}_{Z_p} && \text{for } i = 1 \\ &= * && \text{for } i \neq 1. \end{aligned}$$

Moreover, a short exact sequence of abelian groups $* \to N' \to N \to N'' \to *$ yields (Ch.II, 2.2), up to homotopy a fibration

$$R_\infty K(N', 1) \longrightarrow R_\infty K(N, 1) \longrightarrow R_\infty K(N'', 1)$$

with a long exact sequence

$$\cdots \to \pi_2 R_\infty K(N'', 1) \to \pi_1 R_\infty K(N', 1) \to \pi_1 R_\infty K(N, 1) \to \pi_1 R_\infty K(N'', 1) \to *$$

Thus one can, for N abelian, identify $\pi_{i+1} R_\infty K(N, 1)$ with the i-th left derived functor of the Z_p-completion functor for abelian groups, and the desired result follows easily.

An obvious consequence of 2.2 and Ch.II, 4.8 is:

2.3 Corollary. If N is a nilpotent group and $R = Z_p$, then $R_\infty K(N, 1)$ is a nilpotent space and $\pi_i R_\infty K(N, 1) = *$ for $i \neq 1, 2$.

Now we finally define:

2.4 Ext and Hom completions for nilpotent groups. In view of the above, we can (and will) for every nilpotent group N and $R = Z_p$ (p prime), define

$$\mathrm{Ext}(Z_{p^\infty}, N) = \pi_1 R_\infty K(N, 1)$$

$$\mathrm{Hom}(Z_{p^\infty}, N) = \pi_2 R_\infty K(N, 1).$$

This definition immediately implies that

(i) $\mathrm{Ext}(Z_{p^\infty}, N)$ is nilpotent and the natural map (Ch.IX, §3 and Ch.IV, 2.4)

$$\mathrm{Ext}(Z_{p^\infty}, N) = \pi_1 R_\infty K(N, 1) \longrightarrow \varprojlim \pi_1 R_s K(N, 1) \approx \hat{N}_{Z_p}$$

is onto, with abelian kernel $\varprojlim^1 \pi_2 R_s K(N, 1)$

(ii) $\mathrm{Hom}(Z_{p^\infty}, N)$ is abelian and

$$\mathrm{Hom}(Z_{p^\infty}, N) \approx \varprojlim \pi_2 R_s K(N, 1)$$

and the obvious

2.5 Exactness property. Every short exact sequence of nilpotent groups

$$* \longrightarrow N' \longrightarrow N \longrightarrow N'' \longrightarrow *$$

gives rise to an exact sequence

$$* \longrightarrow \mathrm{Hom}(Z_{p^\infty}, N') \longrightarrow \mathrm{Hom}(Z_{p^\infty}, N) \longrightarrow \mathrm{Hom}(Z_{p^\infty}, N'') \longrightarrow$$

$$\longrightarrow \mathrm{Ext}(Z_{p^\infty}, N') \longrightarrow \mathrm{Ext}(Z_{p^\infty}, N) \longrightarrow \mathrm{Ext}(Z_{p^\infty}, N'') \longrightarrow * .$$

The behavior of the Ext and Hom completions for "ordinary" nilpotent groups is given by:

2.6 Proposition. If N is a nilpotent group, whose p-torsion elements are of bounded order, then

$$\mathrm{Ext}(Z_{p^\infty}, N) \approx \hat{N}_{Z_p} \qquad \text{and} \qquad \mathrm{Hom}(Z_{p^\infty}, N) = * .$$

Proof. For a nilpotent group G, the condition (#) that $\{\pi_2 R_s K(G, 1)\}$ is pro-trivial clearly implies that

$$\mathrm{Ext}(Z_{p^\infty}, G) \approx \hat{G}_{Z_p} \qquad \text{and} \qquad \mathrm{Hom}(Z_{p^\infty}, G) = * .$$

If $* \to G' \to G \to G'' \to *$ is a short exact sequence of nilpotent groups and (#) holds for G' and G'', then (#) holds for G, by Ch.III, 2.5 and 7.1.

By Ch.III, 6.4, (#) holds for Z_p-nilpotent groups, i.e. nilpotent p-torsion groups whose elements are of bounded order. Moreover (#) holds for nilpotent groups without p-torsion, because the upper central series quotients for such groups lack p-torsion [Lazard, Th.3.2] and thus the argument of Ch.IV, 4.4 applies.

The proposition now follows easily since the Malcev completion

$$N \longrightarrow Z[\tfrac{1}{p}] \otimes N$$

has Z_p-nilpotent kernel (Ch.V, 2.8) and has p-torsion free image (Ch.V, 2.7).

We end with a few comments on

2.7 The completion map $N \to \mathrm{Ext}(Z_{p^\infty}, N)$. For a nilpotent group N and $R = Z_p$, the obvious *completion map*

$$N \approx \pi_1 K(N, 1) \xrightarrow{\phi_*} \pi_1 R_\infty K(N, 1) = \mathrm{Ext}(Z_{p^\infty}, N)$$

fits into the commutative *completion triangle*

$$\begin{array}{ccc} & \nearrow & \mathrm{Ext}(Z_{p^\infty}, N) \\ N & & \downarrow \\ & \underset{\phi}{\searrow} & N^{\wedge}_{Z_p} \end{array}$$

Moreover, if N is *abelian*, then this completion map is nothing but the usual *coboundary*

$$\delta\colon\ N \approx \mathrm{Hom}(Z, N) \longrightarrow \mathrm{Ext}(Z_{p^\infty}, N)$$

associated with the obvious short exact sequence

$$* \longrightarrow Z \longrightarrow Z[\tfrac{1}{p}] \longrightarrow Z_{p^\infty} \longrightarrow * \ .$$

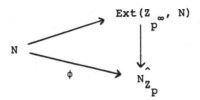

§3. Ext-p-complete nilpotent groups

In this section we will, for a given prime p, discuss the obvious notion of Ext-p-completeness for nilpotent groups and show that

(i) $\mathrm{Ext}(Z_{p^\infty}, N)$ is "really" a completion of N

(ii) $\mathrm{Hom}(Z_{p^\infty}, N)$ is nothing but $\mathrm{Hom}_{\mathrm{groups}}(Z_{p^\infty}, N)$.

3.1 Ext-p-complete nilpotent groups. A nilpotent group N is called Ext-p-complete if

(i) the completion map $N \to \mathrm{Ext}(Z_{p^\infty}, N)$ (2.7) is an isomorphism, and

(ii) $\mathrm{Hom}(Z_{p^\infty}, N) = *$

or equivalently if the space $K(N, 1)$ is Z_p-complete (Ch.I, 5.1).

Then we have the

3.2 Universal property. *If N is a nilpotent group, then*

(i) $\mathrm{Ext}(Z_{p^\infty}, N)$ and $\mathrm{Hom}(Z_{p^\infty}, N)$ are Ext-p-complete, and

(ii) the completion map $N \to \mathrm{Ext}(Z_{p^\infty}, N)$ is universal for homomorphisms from N to Ext-p-complete nilpotent groups.

To prove this we need the following important lemma, which states that $K(N, 1)$ is Z_p-good (Ch.I, 5.1) if N is nilpotent.

3.3 Lemma. *For a nilpotent group N and $R = Z_p$, the map*

$$H_*(K(N, 1); Z_p) \xrightarrow{\phi_*} H_*(R_\infty K(N, 1); Z_p)$$

is an isomorphism.

Proof. Let F be a free abelian group. Then clearly the obvious map

$$F \longrightarrow F^{\wedge}_{Z_p} = \lim_{\leftarrow} F/p^n F$$

is a monomorphism of torsion free abelian groups and induces an isomorphism

$$Z_p \otimes F \approx Z_p \otimes F^{\wedge}_{Z_p} .$$

Thus $F^{\wedge}_{Z_p}/F$ is uniquely p-divisible and, by Ch.IV, 3.3, $K(F^{\wedge}_{Z_p}/F, 1)$ is Z_p-acyclic. And since $R_\infty K(F, 1) \approx K(F^{\wedge}_{Z_p}, 1)$ it follows that the lemma holds for F.

The cases that N is abelian and then nilpotent now follow using Ch.II, 2.2.

Proof of 3.2. Let $R = Z_p$ and consider the obvious diagram of fibrations up to homotopy (Ch.II, 4.8)

$$\begin{array}{ccccc} K(\mathrm{Hom}(Z_{p^\infty}, N), 2) & \longrightarrow & R_\infty K(N, 1) & \longrightarrow & K(\mathrm{Ext}(Z_{p^\infty}, N), 1) \\ \downarrow & & \downarrow & & \downarrow \\ R_\infty K(\mathrm{Hom}(Z_{p^\infty}, N), 2) & \longrightarrow & R_\infty R_\infty K(N, 1) & \longrightarrow & R_\infty K(\mathrm{Ext}(Z_{p^\infty}, N), 1). \end{array}$$

By 3.3 and Ch.II, 5.2, the middle map is a homotopy equivalence, by Ch.I, 6.1, $\pi_i R_\infty K(\mathrm{Hom}(Z_{p^\infty}, N), 2) = *$ for $i < 2$, and, by 2.3, $\pi_i R_\infty K(\mathrm{Ext}(Z_{p^\infty}, N), 1) = *$ for $i > 2$. Hence the outside maps are homotopy equivalences and the groups $\mathrm{Hom}(Z_{p^\infty}, N)$ and $\mathrm{Ext}(Z_{p^\infty}, N)$ are Ext-p-complete.

The universal property now follows easily since the triple structure (Ch.I, 5.6) for R_∞ induces a triple structure for

$$\mathrm{Ext}(Z_{p^\infty}, -): \quad (\text{nilpotent groups}) \longrightarrow (\text{nilpotent groups}).$$

In order to better understand Ext-p-completeness from an algebraic point of view we show:

<u>3.4 First characterization of Ext-p-completeness.</u>

<u>(i) An abelian group N is Ext-p-complete if and only if</u>

$$\mathrm{Hom}(Z[\tfrac{1}{p}], N) = 0 = \mathrm{Ext}(Z[\tfrac{1}{p}], N)$$

<u>(ii) A nilpotent group N is Ext-p-complete if and only if the (abelian) quotients of its upper central series are Ext complete.</u>

<u>Proof.</u> Part (i) follows easily from 2.7 and the "if" part of (ii) is immediate.

Now let N be an Ext complete nilpotent group with center C. Using Ch.IX, 4.1(i), one then can show that in the obvious fibration up to homotopy (Ch.II, 4.8)

$$R_\infty K(N, 1) \longrightarrow R_\infty K(N/C, 1) \longrightarrow R_\infty K(C, 2)$$

the image of

$$\partial: \quad \pi_2 R_\infty K(C, 2) \longrightarrow \pi_1 R_\infty K(N, 1) = N$$

is equal to C. Hence $N/C \approx \pi_1 R_\infty K(N/C, 1)$ and so N/C is Ext-p-complete by 3.2. This easily implies that C is also Ext-p-complete and the "only if" part is now clear.

A more explicit algebraic description of Ext-p-completeness is given in a

3.5 Second characterization of Ext-p-completeness. For a nilpotent group N let

$$L: \quad (N \times N \times N \times \cdots) \longrightarrow (N \times N \times N \times \cdots)$$

denote the function defined by

$$L(x_0, x_1, x_2, \cdots) = (x_0(x_1)^{-p}, x_1(x_2)^{-p}, x_2(x_3)^{-p}, \cdots).$$

If N is abelian, then L is a homomorphism and (Ch.IX, 2.1)

$$\ker L = \varprojlim(N, p) \qquad \operatorname{coker} L = \varprojlim{}^1(N, p)$$

where (N, p) denotes the tower

$$\cdots \xrightarrow{p} N \xrightarrow{p} N \xrightarrow{p} N \longrightarrow * .$$

Thus in the following characterization, part (i) is a special case of part (ii):

(i) An abelian group N is Ext-p-complete if and only if

$$\varprojlim(N, p) = 0 = \varprojlim{}^1(N, p).$$

(ii) A nilpotent group N is Ext-p-complete if and only if the map

$$L: \quad (N \times N \times N \times \cdots) \longrightarrow (N \times N \times N \times \cdots)$$

is a bijection.

Proof. Part (i) follows from 3.4 since [Roos, Th.1]

$$\operatorname{Hom}(Z[\tfrac{1}{p}], N) \simeq \varprojlim(N, p) \qquad \operatorname{Ext}(Z[\tfrac{1}{p}], N) \simeq \varprojlim{}^1(N, p).$$

Part (ii) may be proved by combining 3.4 with the following results for a nilpotent group N with center C:

I. If L is bijective for any two of C, N and N/C, then it is also bijective for the third.

II. If L is bijective for N, then it is also bijective for C.

The proof of I is straightforward, while II follows from the fact that

$$(C \times C \times C \times \cdots) \subset (N \times N \times N \times \cdots)$$

is the subset fixed under the actions

$$(x_0, x_1, x_2, \cdots) \longrightarrow (ux_0u^{-1}, ux_1u^{-1}, ux_2u^{-1}, \cdots) \qquad u \in N$$

and these actions commute with L.

The remainder of this section is devoted to a proof that $\mathrm{Hom}(Z_{p^\infty}, N) \approx \mathrm{Hom}_{\mathrm{groups}}(Z_{p^\infty}, N)$ and for this we need two lemmas with more information on Ext and Hom completions.

3.6 Lemma. For a nilpotent group N,

$$\mathrm{Ext}(Z_{p^\infty}, N) = * \quad \text{if and only if} \quad N \text{ is p-divisible,}$$

(i.e. for each $x \in N$, there is a $y \in N$ with $y^p = x$).

Proof. If $\mathrm{Ext}(Z_{p^\infty}, N) = *$, then in the fibration up to homotopy

$$F \longrightarrow K(N, 1) \xrightarrow{\phi} R_\infty K(N, 1)$$

the fibre F is connected and nilpotent, with (3.1) $\tilde{H}_*(F; Z_p) = 0$. Thus (Ch.V, 3.3) the groups $\pi_i F$ are uniquely p-divisible and the "only if" part readily follows.

If N is p-divisible, then $N \to \text{Ext}(Z_{p^\infty}, N)$ is the trivial map because $\text{Ext}(Z_{p^\infty}, Z[\frac{1}{p}]) = *$ and the "if" part now follows easily, since the functor $\text{Ext}(Z_{p^\infty}, -)$ carries the map $N \to \text{Ext}(Z_{p^\infty}, N)$ to an isomorphism (3.2).

3.7 Lemma. *Let N be a nilpotent group and let $K \subset N$ be the kernel of the completion map* $N \to \text{Ext}(Z_{p^\infty}, N)$. *Then K is the image of the map*

$$\text{Hom}_{\text{groups}}(Z[\tfrac{1}{p}], N) \longrightarrow \text{Hom}_{\text{groups}}(Z, N) = N$$

(i.e. K contains the $x \in N$ which are "infinitely p-divisible in a consistent way") *and moreover*

$$\text{Hom}(Z_{p^\infty}, K) \approx \text{Hom}(Z_{p^\infty}, N).$$

Proof. $\text{Hom}(Z_{p^\infty}, N/K) = *$ since N/K is contained in an Ext-p-complete group (3.2(i)) and thus has a central series whose quotients are subgroups of Ext-p-complete abelian groups (3.4(ii)). Moreover $\text{Ext}(Z_{p^\infty}, N) \to \text{Ext}(Z_{p^\infty}, N/K)$ is an isomorphism, since $\text{Ext}(Z_{p^\infty}, -)$ carries the map $N \to \text{Ext}(Z_{p^\infty}, N)$ to an isomorphism. Thus

$$\text{Hom}(Z_{p^\infty}, K) \approx \text{Hom}(Z_{p^\infty}, N)$$

and $\text{Ext}(Z_{p^\infty}, K) = *$, i.e. (3.6) K is p-divisible. The proposition now follows, because $\text{Ext}(Z_{p^\infty}, Z[\frac{1}{p}]) = *$.

Finally we prove:

3.8 Proposition. For a nilpotent group N, there is a natural isomorphism

$$\mathrm{Hom}(Z_{p^\infty}, N) \approx \mathrm{Hom}_{\mathrm{groups}}(Z_{p^\infty}, N).$$

Proof. Let K be the kernel of the map $N \to \mathrm{Ext}(Z_{p^\infty}, N)$ and let K' be the kernel of the map $K \to Z[\frac{1}{p}] \otimes K$, i.e. K' is the p-torsion subgroup of K (Ch.V, 2.8). Then K/K' is p-torsion free and thus (2.6)

$$\mathrm{Hom}(Z_{p^\infty}, K/K') = * \qquad \mathrm{Hom}(Z_{p^\infty}, K') \approx \mathrm{Hom}(Z_{p^\infty}, K).$$

It is now easy to check that

$$\mathrm{Hom}(Z_{p^\infty}, K') \approx \mathrm{Hom}(Z_{p^\infty}, N)$$

$$\mathrm{Hom}_{\mathrm{groups}}(Z_{p^\infty}, K') \approx \mathrm{Hom}_{\mathrm{groups}}(Z_{p^\infty}, N)$$

and the proposition follows since K' is abelian, as [Kurosh, Vol.II, p.235] every divisible nilpotent torsion group is abelian.

§4. Examples of Ext and Hom completions

We shall give some examples of Ext and Hom completions and review some of Harrison's results on Ext-p-complete abelian groups.

We begin by noting some special cases of 2.6:

4.1 Examples. In each of the following cases N is supposed to be a nilpotent group and will satisfy

$$\mathrm{Ext}(Z_{p^\infty}, N) \approx N^{\wedge}_{Z_p} \qquad \mathrm{Hom}(Z_{p^\infty}, N) = *.$$

(i) N is finitely generated; in this case $N^{\wedge}_{Z_p}$ is the p-profinite completion of N [Serre, p.I-5].

(ii) $N = Z$; in this case $N^{\wedge}_{Z_p} = \underline{\underline{Z}}_p$, where

$$\underline{\underline{Z}}_p = \lim_{\leftarrow} Z/p^n Z$$

denotes the p-adic integers.

(iii) N is Z_p-nilpotent, i.e. there exists an $n < \infty$ such that $x^{p^n} = *$ for all $x \in N$; in this case $N^{\wedge}_{Z_p} = N$.

(iv) N is uniquely p-divisible; in this case $N^{\wedge}_{Z_p} = *$.

Next we give some examples in which the hypotheses of 2.6 are not satisfied:

4.2 Examples.

(i) If $N = Z_{p^\infty}$, then

$$\mathrm{Ext}(Z_{p^\infty}, N) = 0 \qquad \mathrm{Hom}(Z_{p^\infty}, N) = \underline{\underline{Z}}_p.$$

(ii) If $N = Z_p \oplus Z_{p^2} \oplus Z_{p^3} \oplus \cdots$, then N is not Ext-p-complete, $\mathrm{Ext}(Z_{p^\infty}, N)$ is not a torsion group, and $\mathrm{Ext}(Z_{p^\infty}, N) \not\approx N^{\wedge p}_{Z_p}$.

The nature of Ext-p-complete abelian groups is perhaps clarified by:

4.3 Proposition. *An Ext-p-complete abelian group has a canonical $\underline{\underline{Z}}_p$ module structure.*

Proof. For $x \in \underline{\underline{Z}}_p$ and $n \in N$, the product $xn \in N$ is the image of x under

$$\underline{\underline{Z}}_p \approx \mathrm{Ext}(Z_{p^\infty}, Z) \xrightarrow{n_*} \mathrm{Ext}(Z_{p^\infty}, N) \approx N.$$

Equivalently, the module structure of N is given by the Yoneda product

$$\underline{\underline{Z}}_p \otimes N \approx \mathrm{Hom}(Z_{p^\infty}, Z_{p^\infty}) \otimes \mathrm{Ext}(Z_{p^\infty}, N) \longrightarrow \mathrm{Ext}(Z_{p^\infty}, N) \approx N.$$

4.4 Examples.

(i) If N is a finitely generated abelian group, then the natural map

$$\underline{\underline{Z}}_p \otimes N \longrightarrow \mathrm{Ext}(Z_{p^\infty}, N)$$

is an isomorphism.

(ii) The groups

$$\underline{\underline{Z}}_p \otimes \underline{\underline{Z}}_p \qquad \text{and} \qquad \underline{\underline{Z}}_p \otimes Z_{p^\infty} \approx Z_{p^\infty}$$

are not Ext-p-complete, even though they are modules over $\underline{\underline{Z}}_p$.

We conclude with an exposition of the results of [Harrison] on the structure of Ext-p-complete abelian groups (actually we specialized his results on "cotorsion groups" [Harrison, p.370], using the fact that an abelian group is Ext-p-complete if and only if it is a cotorsion group which is uniquely divisible by all primes different from p).

Harrison first analyzes [Harrison, p.373]:

4.5 Torsion free, Ext-p-complete abelian groups. The functors

$$\begin{pmatrix} \text{divisible} \\ \text{p-torsion} \\ \text{abelian groups} \end{pmatrix} \underset{Z_{p^\infty} \otimes -}{\overset{\mathrm{Hom}(Z_{p^\infty}, -)}{\rightleftarrows}} \begin{pmatrix} \text{torsion free} \\ \text{Ext-p-complete} \\ \text{abelian groups} \end{pmatrix}$$

are adjoint equivalences.

Since each divisible p-torsion abelian group can be decomposed as a direct sum of Z_{p^∞}'s [Kurosh, Vol.I, p.165], it follows that a torsion free, Ext-p-complete abelian group N is classified, up to isomorphisms, by the Z_p-dimension of $Z_p \otimes N$.

Next Harrison considers:

4.6 Adjusted Ext-p-complete abelian groups. An Ext-p-complete (with respect to a prime p) abelian group N is called adjusted if N/N_p is divisible, where N_p denote the p-torsion subgroup of N, and one has [Harrison, p.375]:

The functors

$$\begin{pmatrix}\text{p-torsion abelian}\\ \text{groups with no}\\ \text{divisible subgroups}\end{pmatrix} \underset{\operatorname{Tor}(Z_{p^\infty},\, -) = (-)_p}{\overset{\operatorname{Ext}(Z_{p^\infty},\, -)}{\rightleftarrows}} \begin{pmatrix}\text{adjusted}\\ \text{Ext-p-complete}\\ \text{abelian groups}\end{pmatrix}$$

are adjoint equivalences.

Finally Harrison gives [Harrison, p.373]:

4.7 A decomposition of Ext-p-complete abelian groups. For every Ext-p-complete abelian group N, there is a unique splittable short exact sequence

$$* \longrightarrow A \longrightarrow N \longrightarrow F \longrightarrow *$$

such that A is adjusted Ext-p-complete and F is torsion free Ext complete.

In this decomposition

$$A \cong \operatorname{Ext}(Z_{p^\infty}, N_p)$$

$$F \cong \operatorname{Ext}(Z_{p^\infty}, N/N_p)$$

and the splitting is due to [Harrison, p.370]:

4.8 Lemma. If L is an Ext-p-complete abelian group and M is a torsion free abelian group, then Ext(M, L) = 0.

§5. Homotopy and homology properties of the Z_p-completion of a nilpotent space

Our key result is

5.1 Proposition. If $X \in \mathscr{S}_{*N}$ (i.e. X is pointed, connected and Z-nilpotent in the sense of Ch.III, §5), and, of course, $R = Z_p$, then $R_\infty X \in \mathscr{S}_{*N}$ and, for $n \geq 1$, there is a splittable short exact sequence

$$* \longrightarrow \mathrm{Ext}(Z_{p^\infty}, \pi_n X) \longrightarrow \pi_n R_\infty X \longrightarrow \mathrm{Hom}(Z_{p^\infty}, \pi_{n-1} X) \longrightarrow *.$$

Proof. Except for the splittability, this follows by R-completing the Postnikov tower of X, using 2.4 and Ch.II, 4.8. The splittability follows from 4.8, since $\mathrm{Ext}(Z_{p^\infty}, \pi_n X)$ is Ext-p-complete and $\mathrm{Hom}(Z_{p^\infty}, \pi_{n-1} X)$ is torsion free.

An important case of 5.1 is:

5.2 Example. If the groups $\pi_n X$ are all finitely generated abelian (and $X \in \mathscr{S}_{*N}$), then

$$\pi_n R_\infty X \approx \underline{\underline{Z}}_p \otimes \pi_n X$$

where $\underline{\underline{Z}}_p$ denotes the p-adic integers (see 4.1); and of course

$$\underline{\underline{Z}}_p \otimes Z \approx \underline{\underline{Z}}_p$$

$$\underline{\underline{Z}}_p \otimes Z_{p^j} \approx Z_{p^j}$$

$$\underline{\underline{Z}}_p \otimes Z_{q^j} \approx 0 \qquad \text{for any prime } q \neq p.$$

Another easy consequence of 5.1 and Ch.I, 5.2 is:

5.3 Proposition. If $X \in \mathscr{S}_{*N}$, then

(i) X is R-good (i.e. $\tilde{H}_*(X; Z_p) \approx \tilde{H}_*(R_\infty X; Z_p)$),

(ii) $R_\infty X$ is R-complete (i.e. $R_\infty X \simeq R_\infty R_\infty X$).

And, together with Ch.II, 2.8, the above results imply:

5.4 Proposition. For a space $X \in \mathscr{S}_{*N}$, the following are equivalent:

(i) X is Z_p-complete.

(ii) The groups $\pi_n X$ are Ext-p-complete.

(iii) Whenever a map $f: K \to L \in \mathscr{S}_*$ induces an isomorphism $H_*(K; Z_p) \approx H_*(L; Z_p)$, then it induces a bijection $[L, X] \approx [K, X]$ of pointed homotopy classes of maps (Ch.VIII, §4).

5.5 Example. For $R = Z_p$ and $n \geq 1$, there is an isomorphism

$$[R_\infty S^n, R_\infty S^n] \approx \pi_n R_\infty S^n \approx \underline{\underline{Z}}_p$$

which assigns to each map $f: R_\infty S^n \to R_\infty S^n$ the obvious degree $\deg f \in \underline{\underline{Z}}_p$.

We end with a brief discussion of

5.6 The homology of $R_\infty X$. Let $X \in \mathscr{S}_{*N}$, let $R = Z_p$ and let q be a prime. Then

$$\tilde{H}_*(R_\infty X; Z_q) \approx \tilde{H}_*(X; Z_q) \quad \text{if } q = p$$

$$= 0 \quad \text{if } q \neq p.$$

<u>Proof</u>. The case $q = p$ is just 5.3, and the case $q \neq p$ follows from Ch.V, 3.3, because the groups $\pi_n R_\infty X$ are uniquely q-divisible (An easy argument, using 3.4, shows that <u>all nilpotent groups which are Ext-p-complete are uniquely q-divisible for all primes $q \neq p$</u>).

<u>5.7 Remark</u>. If $X \in \mathscr{S}_{*N}$ and $R = Z_p$ the <u>integral</u> homology $H_*(R_\infty X; Z)$ is uniquely q-divisible for primes $q \neq p$, but still not very well behaved. For example, if S^n is an odd sphere and $k \geq 2$, then $H_{kn}(R_\infty S^n; Z)$ is a huge Q-module (because the Q-completion of $R_\infty S^n$ has the homotopy type of $K(Q \otimes \underline{\underline{Z}}_p, n)$ and $Q \otimes \underline{\underline{Z}}_p$ has uncountable dimension over Q).

§6. p-completions of nilpotent spaces

We introduce a notion of p-completion for nilpotent spaces which is merely an "up to homotopy" version of our Z_p-completion and which generalizes the usual p-profinite completion [Quillen (PG)] [Sullivan, Ch.3] for simply connected spaces of finite type. We also consider various preservation properties of p-completions and observe that the p-completion factors through the $Z_{(p)}$-localization of Chapter V.

6.1 p-completions. For $X \in \mathcal{S}_{*N}$, a p-completion of X is a map $X \to \bar{X} \in \mathcal{S}_{*N}$ such that

(i) $\bar{X}$ is Z_p-complete, and

(ii) the induced map $\tilde{H}_*(X; Z_p) \to \tilde{H}_*(\bar{X}; Z_p)$ is an isomorphism.

The results of §5 then immediately imply:

6.2 Proposition. p-completion is well-defined and functorial on the pointed homotopy category of nilpotent spaces. It is induced by the functor R_∞ where $R = Z_p$.

6.3 Homotopy characterization of $R_\infty X$ $(R = Z_p)$. For $X \in \mathcal{S}_{*N}$, the Z_p-completion $X \to R_\infty X$ is a p-completion, and, in the pointed homotopy category, any p-completion $X \to \bar{X}$ is canonically equivalent to $X \to R_\infty X$.

6.4 Example. There are Z_p-homology equivalences

$$K(Z_{p^\infty}, n-1) \longrightarrow K(Z, n) \longrightarrow K(Z_{(p)}, n) \longrightarrow K(\hat{Z}_p, n)$$

and thus the Z_p-complete space $K(\hat{Z}_p, n)$ is a p-completion of all these spaces.

Next we discuss the "preservation properties" of p-completions. We already have from Ch.II, §4:

6.5 Proposition.

(i) If $p\colon E \to B \in \mathcal{S}_{*N}$ *is a fibration with* $\pi_1 E \to \pi_1 B$ *onto, then* $R_\infty p\colon R_\infty E \to R_\infty B$ *is a fibration with* $p^{-1}* \in \mathcal{S}_{*N}$ *and* $R_\infty(p^{-1}*) \simeq (R_\infty p)^{-1}*$.

(ii) If $X \in \mathcal{S}_{*N}$ *is 1-connected, then* $R_\infty \Omega X \simeq \Omega R_\infty X$, where Ω denotes the loop functor [May, p.99].

(iii) If $X, Y \in \mathcal{S}_{*N}$, *then* $R_\infty(X \times Y) \simeq R_\infty X \times R_\infty Y$.

Because all spaces in $\mathcal{S}_{*N}$ are Z_p-good, one can apply Ch.I, 5.5 to show:

6.6 Proposition.

(i) If $i\colon A \to X \in \mathcal{S}_{*N}$ *is a cofibration, then* $R_\infty i\colon R_\infty A \to R_\infty X$ *is a cofibration and* $R_\infty(X/A) \simeq R_\infty(R_\infty X/R_\infty A)$.

(ii) If $X \in \mathcal{S}_{*N}$, *then* $R_\infty SX \simeq R_\infty S R_\infty X$, where S denotes the suspension [May, p.124].

(iii) If $X, Y \in \mathcal{S}_{*N}$, *then* $R_\infty(X\ Y) \simeq R_\infty(R_\infty X \vee R_\infty Y)$.

(iv) If $X, Y \in \mathcal{S}_{*N}$, *then* $R_\infty(X \wedge Y) \simeq R_\infty(R_\infty X \wedge R_\infty Y)$.

Of course the "preservation properties" in 6.6 are not as pleasant as one might hope. This is because the usual "direct limit constructions" do not preserve Z_p-completeness for nilpotent spaces.

6.7 Example. The spaces $R_\infty S^m \vee R_\infty S^n$ for $m, n \geq 2$, $R_\infty S^m \wedge R_\infty S^n$ for $m, n \geq 1$ and $S^2 R_\infty S^m$ for m odd, are *not* Z_p-complete because the groups

$$\pi_{m+n-1}(R_\infty S^m \vee R_\infty S^n) \approx \pi_{m+n-1}R_\infty S^m \oplus \pi_{m+n-1}R_\infty S^n \oplus \underline{\underline{Z}}_p \otimes \underline{\underline{Z}}_p$$

$$\pi_{m+n}(R_\infty S^m \wedge R_\infty S^n) \approx \underline{\underline{Z}}_p \otimes \underline{\underline{Z}}_p$$

are not Ext-p-complete and because

$$H_*(R_\infty S^m; Q) \not\approx H_{*+2}(R_\infty S^{m+2}; Q)$$

for m odd (see 5.7).

We conclude by

6.8 Factoring p-completions through $Z_{(p)}$-localizations. Let $Z_{(p)} \subset Q$ denote the integers localized at p. Then, for $X \in \mathcal{S}_{*N}$, the map

$$\tilde{H}_*(X; Z_p) \longrightarrow \tilde{H}_*((Z_{(p)})_\infty X; Z_p)$$

is an isomorphism (Ch.V, 3.2), and hence the map

$$(Z_p)_\infty X \longrightarrow (Z_p)_\infty (Z_{(p)})_\infty X \qquad \in \mathcal{S}_{*N}$$

is a homotopy equivalence. Moreover (5.6) the Z_p-complete spaces in $\mathcal{S}_{*N}$ are all $Z_{(p)}$-complete.

Thus, up to homotopy, the Z_p-completion on $\mathcal{S}_{*N}$ can be viewed as a two-step process:

(i) $Z_{(p)}$-completion for nilpotent spaces, followed by

(ii) Z_p-completion for $Z_{(p)}$-complete spaces.

§7. *p-completions of function spaces*

In preparation for the arithmetic square fracture lemma (§8) we show here that the homotopy types of the *pointed function spaces* (Ch.VIII, §4)

$$\hom_*(W, X) \qquad \text{and} \qquad \hom_*(W, R_\infty X)$$

$(R = Z_p)$ are often closely related. The results and proofs are very similar to those for localizations (Ch.V, §5), except that we assume that X is not only *nilpotent*, but also has *finitely generated (nilpotent) homotopy groups* (see 7.4 and 8.5). We shall implicitly use the fact that *the functor* $\operatorname{Ext}(Z_{p^\infty}, -)$ *preserves exact sequences of such groups*. We also remind the reader that, *for a finitely generated nilpotent group* N

$$\operatorname{Ext}(Z_{p^\infty}, N) \approx \hat{N}_{Z_p} \approx \text{\underline{the p-profinite completion of}}\ N$$

and that, *for a finitely generated abelian group* N

$$\operatorname{Ext}(Z_{p^\infty}, N) \approx \underline{\underline{Z}}_p \otimes N$$

We start with a proposition which implies that, under suitable conditions, the Z_p-completion of any component of $\hom_*(W, X)$ has the same homotopy type as the corresponding component of $\hom_*(W, R_\infty X)$, where $R = Z_p$.

7.1 Proposition. *Let* $X \in \mathcal{S}_{*N}$ *be fibrant and have finitely generated nilpotent homotopy groups, let* $W \in \mathcal{S}_{*C}$ *be finite and let* $R = Z_p$. *Then, for every map* $f\colon W \to X \in \mathcal{S}_{*C}$ *and all* $i \geq 1$, *(i)* $\pi_i(\hom_*(W, X), f)$ *is finitely generated nilpotent,*

(ii) $\pi_i(\hom_*(W, R_\infty X), \phi f)$ *is Ext-p-complete nilpotent, and*

(iii) the map $\phi\colon X \to R_\infty X$ *induces an isomorphism*

$$\pi_i(\hom_*(W, X), f)^{\wedge}_{Z_p} \approx \pi_i(\hom_*(W, R_\infty X), \phi f).$$

The proof is similar to that of Ch.V, 5.1.

Again, the relation between the sets of components of $\hom_*(W, X)$ and $\hom_*(W, R_\infty X)$, i.e. the relation between the *pointed homotopy classes of maps* (Ch.VIII, §4)

$$[W, X] \qquad \text{and} \qquad [W, R_\infty X]$$

is not so easy to describe. Of course, as in Ch.V, 5.3, one proves:

7.2 Proposition. *Let* $X \in \mathcal{S}_{*N}$ *be fibrant and have finitely generated nilpotent homotopy groups, let* $W \in \mathcal{S}_{*C}$ *and let either* W *be a reduced suspension [May, p.124] or* X *be a homotopy associative H-space (Ch.I, 7.5). Then*

(i) $[W, X]$ *is a finitely generated nilpotent group,*

(ii) $[W, R_\infty X]$ *is an Ext-p-complete nilpotent group, and*

(iii) the map $\phi\colon X \to R_\infty X$ *induces an isomorphism*

$$[W, X]^{\wedge}_{Z_p} \approx [W, R_\infty X].$$

In general, however, the sets $[W, X]$ and $[W, R_\infty X]$ do *not* come with a group structure. Still, as in Ch.V, 5.5, one can prove a useful result for *neighborhood groups*:

7.3 Proposition. *Let* $X \in \mathcal{S}_{*N}$ *be fibrant and have finitely generated nilpotent homotopy groups and let* $W \in \mathcal{S}_{*C}$ *be finite and* reduced. Then, for every map $f: W \to X \in \mathcal{J}_{*C}$,

(i) $[W, R_\infty X]_{\phi f}$ is an Ext-p-complete nilpotent group, and

(ii) the map $\phi: X \to R_\infty X$ induces an isomorphism

$$([W, X]_f)\hat{}_{Z_p} \approx [W, R_\infty X]_{\phi f} .$$

7.4 Remark. In proposition 7.1, 7.2 and 7.3 we have supposed that X has finitely generated nilpotent homotopy groups, and although this condition can not be omitted (8.5), it should be noted that propositions 7.1, 7.2 and 7.3 remain true if "finitely generated nilpotent" is everywhere replaced by "$Z[J^{-1}]$-nilpotent and finitely generated over $Z[J^{-1}]$, where J is a fixed set of primes and a $Z[J^{-1}]$-nilpotent group is called finitely generated over $Z[J^{-1}]$ if it has a central series whose (abelian) quotients are finitely generated $Z[J^{-1}]$-modules in the usual sense.

§8. The arithmetic square fracture lemma

We end this chapter with a <u>fracture lemma</u> involving Z_p-<u>completions</u>, which is essentially due to Sullivan. It will be formulated in terms of the <u>arithmetic square</u> [Sullivan, 3.58], i.e., in the notation of Ch.V, 6.1, a diagram of the form

$$\begin{array}{ccc} X_{(I)} & \longrightarrow & \prod_{p \in I} (Z_p)_\infty X \\ \downarrow & & \downarrow \\ X_{(0)} & \longrightarrow & (\prod_{p \in I} (Z_p)_\infty X)_{(0)} \end{array}$$

where $X \in \mathcal{S}_{*N}$, I is a set of primes, the top map is induced by the Z_p-completions and the bottom map is the Q-completion of the top map.

The main result is the

<u>8.1 Arithmetic square fracture lemma</u>. <u>Let</u> $X \in \mathcal{S}_{*N}$ <u>have finitely generated homotopy groups, let</u> $W \in \mathcal{S}_{*C}$ <u>be finite and let</u> I <u>be a set of primes. Then</u>

<u>(i) the arithmetic square</u>

$$\begin{array}{ccc} X_{(I)} & \longrightarrow & \prod_{p \in I} (Z_p)_\infty X \\ \downarrow & & \downarrow \\ X_{(0)} & \longrightarrow & (\prod_{p \in I} (Z_p)_\infty X)_{(0)} \end{array}$$

<u>is, up to homotopy, a fibre square, and</u>

<u>(ii) the induced square of pointed homotopy classes of maps (Ch.VIII, §4)</u>

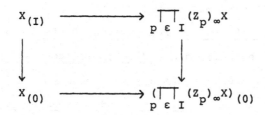

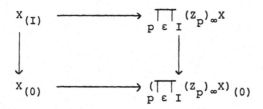

$$\begin{array}{ccc}
[W, X_{(I)}] & \longrightarrow & [W, \prod_{p \in I} (Z_p)_\infty X] \\
\downarrow & & \downarrow \\
[W, X_{(0)}] & \longrightarrow & [W, (\prod_{p \in I} (Z_p)_\infty X)_{(0)}]
\end{array}$$

is a pull-back in which the upper map is an injection.

Proof. The proof is essentially the same as that of the prime fracture lemma (Ch.V, 6.2) and uses 7.3, 7.4, and the following *group theoretic analogue* of 8.1 (whose proof is similar to that of Ch.V, 6.5).

8.2 Lemma. *If N is a finitely generated nilpotent group and I is a set of primes, then the natural diagram*

$$\begin{array}{ccc}
Z_{(I)} \otimes N & \longrightarrow & \prod_{p \in I} N^{\wedge}_{Z_p} \\
\downarrow & & \downarrow \\
Q \otimes N & \longrightarrow & Q \otimes (\prod_{p \in I} N^{\wedge}_{Z_p})
\end{array}$$

is a pull-back in which the top map is an injection. Moreover, every element $u \in Q \otimes (\prod_{p \in I} N^{\wedge}_{Z_p})$ *can be expressed as* $u = vw$, *where* v *(resp.* w*) is in the image of* $Q \otimes N$ *(resp.* $\prod_{p \in I} N^{\wedge}_{Z_p}$*).*

8.3 Remark. The arithmetic square fracture lemma shows that a space $X \in \mathscr{S}_{*N}$ with *finitely generated homotopy groups* is, up to homotopy, determined by its various Z_p-completions together with "rational information". However its most interesting feature is the assertion that, *for* $W \in \mathscr{S}_{*C}$ *finite, the map*

$$[W, X_{(I)}] \longrightarrow \prod_{p \in I} [W, (Z_p)_\infty X]$$

is an injection. Since (6.8) $(Z_p)_\infty X \simeq (Z_p)_\infty X_{(p)}$, this is stronger than the previous result (Ch.V, 6.2) that

$$[W, X_{(I)}] \longrightarrow \prod_{p \in I} [W, X_{(p)}]$$

is an injection.

8.4 A relation between Z_p-completions and $Z_{(p)}$-completions. We showed in 6.8 that, for $X \in \mathcal{S}_{*N}$, the Z_p-completion $(Z_p)_\infty X$ is, up to homotopy, determined by the $Z_{(p)}$-completion $X_{(p)}$.

On the other hand 8.1 implies that, for $X \in \mathcal{S}_{*N}$ with finitely generated homotopy groups, the homotopy type of $X_{(p)}$ is determined by $(Z_p)_\infty X$ and the rational information of the, up to homotopy, fibre square

$$\begin{array}{ccc} X_{(p)} & \longrightarrow & (Z_p)_\infty X \\ \downarrow & & \downarrow \\ X_{(0)} & \longrightarrow & ((Z_p)_\infty X)_{(0)} \end{array} .$$

8.5 The restrictions on W and X. The condition in 8.1 that $W \in \mathcal{S}_{*C}$ be finite is clearly unnecessarily restrictive and can be relaxed as in Ch.V, §7.

However, the restriction that $X \in \mathcal{S}_{*N}$ have finitely generated homotopy groups cannot so easily be dropped, although it can be modified as in 7.4. Some of the difficulties that arise are, for instance:

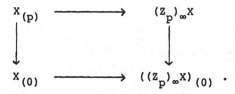

(i) For $X = K(Z_{p^\infty}, n)$, the Z_p-completions of the components of $\hom_*(S^n, X)$ are not homotopy equivalent to the corresponding components of $\hom_*(S^n, (Z_p)_\infty X)$.

(ii) For $X = K(Z_{p^\infty}, n)$, the map

$$[S^n, X] \longrightarrow \prod_{p \text{ prime}} [S^n, (Z_p)_\infty X]$$

is <u>not</u> an injection.

(iii) The arithmetic square is <u>not</u>, up to homotopy a fibre square for

$$X = K(\Sigma_p Z_p, n)$$

where p runs over all primes.

We end with an interesting consequence of 8.1:

<u>8.6 Proposition</u>. <u>Let</u> $X \in \mathcal{S}_{*N}$ <u>be fibrant and have finitely generated homotopy groups, let</u> $W \in \mathcal{S}_{*C}$ <u>be finite, let</u> $f, g: W \to X \in \mathcal{S}_{*C}$ <u>be such that</u>

$$[f] \neq [g] \qquad \in [W, X]$$

<u>and let p be a prime. Then there exists a map</u> $u: X \to Y \in \mathcal{S}_{*N}$ <u>such that</u>

$$[uf] \neq [ug] \qquad \in [W, Y]$$

<u>and such that each</u> $\pi_i Y$ <u>is a finite p-group</u>.

<u>Proof</u>. Let $R = Z_p$. Then (Ch.V, 7.5) one readily sees, by inspecting the E_1-term of the extended homotopy spectral sequence of $\{R_s X\}$ (Ch.I, 4.4), that each $\pi_i R_s X$ is a finite p-group for $s < \infty$. In view of Ch.IX, §3 this implies

$$[W, R_\infty X] \approx \varprojlim [W, R_s X]$$

and so the proposition follows from 8.1.

§9. Curtis convergence theorems

We end this chapter with some best possible convergence results for the homotopy spectral sequence $\{E_r(X; Z_p)\}$ of a nilpotent space X (Ch.I, 4.4) and observe at the end of this section that, as in Ch.IV, 5.6, these results readily extend to the lower p-central series spectral sequences. We also indicate a generalization to certain not necessarily nilpotent spaces. Similar convergence results for $R \subset Q$ were obtained in Ch.IV, 5.6 and Ch.V, 3.7.

The main result of this section (9.1) was proved initially in [Bousfield-Kan (HS), §6] for simply connected spaces, by combining Rector's variation [Rector (AS)] on Curtis' fundamental theorem (Ch.IV, 5.6 and [Curtis (H)]) with some ad-hoc simplicial arguments. Our present approach, however, is more direct and yields stronger results.

9.1 Curtis convergence theorem for $R = Z_p$. If $X \in \mathscr{S}_{*N}$ is such that in each $\pi_i X$ the p-torsion elements are of bounded order, then $\{E_r(X; Z_p)\}$ converges to $\pi_* X$ in the following sense:

(i) $\{E_r(X; Z_p)\}$ is Mittag-Leffler (Ch.IX, 5.5) in all dimensions ≥ 1.

(ii) For each $i \geq 1$ there exists a natural filtration

$$\cdots \subset F^2\pi_i X \subset F^1\pi_i X \subset F^0\pi_i X = \pi_i X$$

such that

$$(F^s/F^{s+1})\pi_i X \approx E_\infty^{s,s+i}(X; Z_p) \qquad \text{for } s \geq 0$$

and such that $\bigcap_s F^s\pi_i X$ is the kernel of the completion map (Ch.VI, 2.6 and 2.7)

$$\pi_i X \longrightarrow \mathrm{Ext}(Z_{p^\infty}, \pi_i X) \approx (\pi_i X)\hat{}_{Z_p}$$

or equivalently (Ch.VI, 3.7) is the image of the map

$$\mathrm{Hom}_{\text{groups}}(Z[\tfrac{1}{p}], \pi_i X) \longrightarrow \mathrm{Hom}_{\text{groups}}(Z, \pi_i X) \approx \pi_i X$$

(i.e. $\bigcap_s F^s \pi_i X$ consists of the elements of $\pi_i X$ which are "infinitely p-divisible in a consistent way").

We remark that [Kurosh, Vol.I, p.173], for $\pi_i X$ abelian, the p-torsion elements in $\pi_i X$ are of bounded order if and only if the p-torsion subgroup of $\pi_i X$ decomposes as a direct sum of copies of Z_{p^j} where the j are bounded.

The result of 9.1 is essentially best possible because:

9.2 Proposition. For G abelian and $n \geq 1$, the spectral sequence $\{E_r(K(G, n); Z_p)\}$ is Mittag-Leffler in all dimensions ≥ 1 if and only if the p-torsion subgroup of G decomposes as a (possibly trivial) direct sum of copies of Z_{p^∞} and copies of Z_{p^j} where the j are bounded.

Of course, if G has Z_{p^∞} as a direct summand, then the condition 9.1(ii) will not hold for $X = K(G, n)$.

9.3 Remark. Theorem 9.1 is not the most general convergence theorem. For instance, it is clear that:

If $X \in \mathcal{S}_{*c}$ is such that $H_i(X; Z_p)$ is finite for each $i \geq 1$, then each of the groups $E_r^{s,t}(X; Z_p)$ is also finite. Consequently $\{E_r(X; Z_p)\}$ is Mittag-Leffler in all dimensions ≥ 1 (Ch.IX, 5.5)

<u>and converges completely to</u> $\pi_*(Z_p)_\infty X$ (which may be very different from π_*X).

<u>Proof of 9.1</u>. The proof of Ch.VI, 2.6 implies that the tower of groups $\{\pi_q R_s K(\pi_i X, i)\}$ is pro-trivial for $i \geq 1$ and $q \neq i$. Thus Ch.II, 5.4 and Ch.III, 7.1 show that the obvious tower maps

$$\{\pi_i R_s K(\pi_i X, i)\} \longrightarrow \{\pi_i R_s X^{(i)}\} \longleftarrow \{\pi_i R_s X\}$$

are pro-isomorphisms for $i \geq 1$, where $X^{(i)}$ denotes the i-th Postnikov space of X [May, p.31]. Using Ch.IV, 2.4 and Ch.VI, 2.3 and 3.7, it is now easy to show that the obvious maps

$$\pi_i X \longrightarrow \pi_i R_s K(\pi_i X, i) \qquad i \geq 1, \quad 0 \leq s < \infty$$

are onto and that

$$\bigcap_{0 \leq s < \infty} \ker(\pi_i X \longrightarrow \pi_i R_s K(\pi_i X, i))$$

is the subgroup of $\pi_i X$ consisting of the elements of $\pi_i X$ which are "infinitely p-divisible in a consistent way". The theorem now follows easily.

<u>Proof of 9.2</u>. By [Bousfield-Kan (HS), §15]

$$\begin{aligned} E_2^{s,t}(K(G, n); Z_p) &\approx G \otimes Z_p && \text{for } s = 0,\ t = n \\ &\approx \operatorname{coker}(G, Z_p) && \text{for } s > 0,\ t-s = n \\ &\approx \ker(G, Z_p) && \text{for } s \geq 0,\ t-s = n+1 \\ &= 0 && \text{otherwise} \end{aligned}$$

where coker(G, Z_p) and ker(G, Z_p) denote the cokernel and kernel of the obvious composition

$$\mathrm{Tor}(G, Z_p) \longrightarrow G \longrightarrow G \otimes Z_p.$$

Moreover, for $r \geq 2$, $E_r(K(G, n); Z_p)$ suspends isomorphically to the E_r-term of the Adams spectral sequence for the $K(G)$ spectrum. Thus, if $r \geq 2$ and $t-s = n+1$, then $E_r^{s,t}(K(G, n); Z_p)$ is isomorphic to the kernel of the obvious composition

$$\mathrm{Tor}(G, Z_p) \longrightarrow G \longrightarrow G \otimes Z_{p^{r-1}}.$$

Now it is easy to show that $\{E_r(K(G, n); Z_p)\}$ is Mittag-Leffler in all dimensions ≥ 1 if and only if the p-torsion subgroup of G decomposes as $I \oplus M$, where I is divisible and $p^k M = 0$ for some k. The proposition then follows from [Kurosh, Vol.I, p.165 and p.173].

9.4 Remark. The analogue of theorem 9.1 for the lower p-central series spectral sequences ([Rector (AS)], [Quillen (PG)] and [Bousfield-Curtis]) can be proved in the same way as 9.1. The main reason for this is that:

(i) for towers of groups, the Mittag-Leffler property (Ch.IX, 3.5) is a pro-isomorphism (Ch.III, 2.1) invariant, and thus (Ch.IX, 5.6)

(ii) for spectral sequences of towers of fibrations in $\mathcal{S}_{*C}$, the Mittag-Leffler property in dimension i is a weak pro-homotopy equivalence (Ch.III, 3.1) invariant.

Note that complete convergence (Ch.IX, 5.3) is, in general, not invariant under weak pro-homotopy equivalences. For example, if $\underline{\underline{Z}}_p$ denotes the p-adic integers, then there is a weak pro-homotopy equivalence

$$\begin{array}{ccccccccc}
\cdots & \xrightarrow{f} & K(\underline{\underline{Z}}_p, n) & \xrightarrow{f} & K(\underline{\underline{Z}}_p, n) & \xrightarrow{f} & K(\underline{\underline{Z}}_p, n) & \longrightarrow & * \\
 & & \downarrow f & & \downarrow f & & \downarrow g & & \downarrow \\
\cdots & \xrightarrow{f} & K(\underline{\underline{Z}}_p, n) & \xrightarrow{f} & K(\underline{\underline{Z}}_p, n) & \xrightarrow{g} & K(\underline{\underline{Z}}_p/Z, n) & \longrightarrow & *
\end{array}$$

$(n \geq 1)$, where f induces $\underline{\underline{Z}}_p \xrightarrow{p} \underline{\underline{Z}}_p$ and g induces the quotient map $\underline{\underline{Z}}_p \longrightarrow \underline{\underline{Z}}_p/Z$. However the spectral sequence of the upper tower converges completely, while that of the lower tower does not.

Chapter VII. A glimpse at the R-completion of non-nilpotent spaces

§1. Introduction

Although the R-completion is quite well understood for nilpotent spaces (Ch.V and Ch.VI), the situation for non-nilpotent spaces is still very mysterious. So far we have essentially dealt with only one non-nilpotent example, in Ch.IV, 5.3, where we showed that for any free group F

$$R_\infty K(F, 1) \simeq K(\hat{F_R}, 1)$$

and our main purpose in this chapter is to discuss some other non-nilpotent spaces and indicate how little is known about them and how much more work remains to be done for non-nilpotent spaces.

§2 This first section contains some easy homotopy characterizations of the R-completion for R-good spaces.

§3, §4 and §5 form the central part of this chapter; we discuss various non-nilpotent spaces and show, in particular, that:

(i) any space X with an R-perfect fundamental group (i.e. $H_1(X; R) = 0$) is R-good for $R \subset Q$ and $R = Z_p$,

(ii) any space with finite homotopy groups in each dimension is R-good for $R \subset Q$ and $R = Z_p$,

(iii) any space with a finite fundamental group is Z_p-good for all primes p, but

(iv) the projective plane as well as some finite wedge of circles is not Z-good.

§6 This last section contains some comments on possible R-homotopy theories for $R \subset Q$ and $R = Z_p$.

Notation. In this chapter we will mainly work in the category $\mathcal{S}_{*c}$ of pointed connected spaces.

Throughout this chapter the ring R will be $R \subset Q$ or $R = Z_p$, except in §2, where we allow arbitrary (solid) rings.

§2. Homotopy characterizations of the R-completion for R-good spaces

For any (solid) ring R we

(i) give universal properties which characterize, up to homotopy, the R-completion for R-good spaces, and

(ii) formulate an "up to homotopy" version of the R-completion for R-good spaces, which generalizes the localizations (Ch.V, §4) and the p-completions (Ch.VI, §6); a generality which is justified because there are many R-good spaces which are not nilpotent (see §3, §4 and §5).

First we have (in view of Ch.I, §5 and Ch.II, 2.8) the

2.1 Universal properties. For an R-good space $X \in \mathcal{S}_{*C}$, the map $\phi: X \to R_\infty X$ has the following "up to homotopy" universal properties:

(i) $\phi: X \to R_\infty X$ is terminal among the maps $f: X \to Y \in \mathcal{S}_{*C}$ for which $f_*: H_*(X; R) \simeq H_*(Y; R)$, i.e. for any such map f, there exists a unique homotopy class of maps $u: Y \to R_\infty X \in \mathcal{S}_{*C}$ such that $uf \simeq \phi$.

(ii) $\phi: X \to R_\infty X$ is initial among the maps $f: X \to Y \in \mathcal{S}_{*C}$ for which Y is R-complete and fibrant, i.e. for any such map f, there exists a unique homotopy class of maps $u: R_\infty X \to Y \in \mathcal{S}_{*C}$ such that $u\phi \simeq f$.

Next we introduce our "up to homotopy" version of the R-completions, which we call

2.2 Semi-R-completions. For this we first say that a space $K \in \mathcal{S}_{*C}$ is *semi-R-complete* if it is fibrant and every map $f: X \to Y \in \mathcal{S}_{*C}$ which induces an isomorphism $H_*(X; R) \approx H_*(Y; R)$, also induces a bijection of the pointed homotopy classes of maps $[Y, K] \approx [X, K]$. This is motivated by Ch.II, 2.8, which implies that, *for every* $W \in \mathcal{S}_{*C}$, *the R-completion* $R_\infty W$ *is semi-R-complete*.

For $X \in \mathcal{S}_{*C}$, *a semi-R-completion of* X now is a map $X \to \overline{X} \in \mathcal{S}_{*C}$ such that

(i) $\overline{X}$ is semi-R-complete, and

(ii) the induced map $H_*(X; R) \to H_*(\overline{X}; R)$ is an isomorphism.

Although, in general, the R-completion of X need *not* be a semi-R-completion of X, one clearly has:

2.3 Proposition. *The semi-R-completion is well-defined and functorial on the pointed homotopy category of R-good spaces. It is induced by the functor* R_∞.

2.4 Homotopy characterization of $R_\infty X$. *For an R-good space* $X \in \mathcal{S}_{*C}$, *the R-completion* $\phi: X \to R_\infty X$ *is a semi-R-completion, and, in the pointed homotopy category, any semi-R-completion of* X *is canonically equivalent to* $\phi: X \to R_\infty X$.

Note that each "R-homology type" of R-good spaces contains exactly one homotopy type of (semi-) R-complete spaces; and the (semi-) R-completion "selects" that homotopy type.

§3. Spaces with an R-perfect fundamental group

An interesting class of spaces which are R-good consists of the spaces whose fundamental group is R-perfect. We will show in several examples that, for such spaces, $\pi_* R_\infty X$ may be very different from $\pi_* X$. We start with recalling the definition of

3.1 R-perfect groups. A group G is called R-perfect if $H_1(G; R) = 0$, i.e. if

$$R \otimes (\text{abelianization of } G) = 0.$$

Thus a Z-perfect group is nothing but a group which is perfect in the usual sense. Clearly every perfect group is R-perfect.

An immediate consequence of this definition, Ch.I, 5.2 and 6.1, Ch.V, 3.4 and Ch.VI, 5.3 is

3.2 Proposition. Let $X \in \mathscr{S}_{*C}$, let $R \subset Q$ or $R = Z_p$ (p prime) and let $\pi_1 X$ be R-perfect. Then X is R-good and $R_\infty X$ is simply connected.

3.3 Examples.

(i) The projective plane P^2 is Z_p-good for $p \neq 2$; actually P^2 is also Z_2-good, but this we will only see in §5.

(ii) Let A_∞ denote the infinite alternating group, i.e.

$$A_\infty = \lim_{\rightarrow} A_n$$

where A_n denotes the alternating group of degree n, i.e. the group of the even permutations of $\{1,\cdots,n\}$. Then $K(A_\infty, 1)$ is R-good

<u>for $R \subset Q$ and</u> $R = Z_p$, because [Kurosh, Vol.I, p.68] A_n is simple for $n \geq 5$ and thus A_∞ is perfect.

(iii) Let S_∞ denote the infinite symmetric group, i.e.

$$S_\infty = \varinjlim S_n$$

where S_n denotes the symmetric group of degree n, i.e. the group of the permutations of $\{1,\cdots,n\}$. Then $K(S_\infty, 1)$ <u>is R-good for $R \subset Q$ and</u> $R = Z_p$, in spite of the fact that S_∞ is only Z_p-perfect for $p \neq 2$. To prove this one observes that there is, for each n, an obvious monomorphism $A_n \times Z_2 \to S_{n+2}$ which is compatible with the inclusion $A_n \to S_{n+2}$ and the projection $S_{n+2} \to Z_2$. As furthermore

$$H_*(A_\infty, Z) = \varinjlim H_*(A_n, Z)$$

it is not hard to see that, in the fibration

$$K(A_\infty, 1) \longrightarrow K(S_\infty, 1) \longrightarrow K(Z_2, 1),$$

$\pi_1 K(Z_2, 1) = Z_2$ acts trivially on $H_*(K(A_\infty, 1), Z) = H_*(A_\infty, Z)$. The desired result now follows from (ii) and Ch.II, 5.1.

The last two of these examples are $K(\pi, 1)$'s whose Z-completion has as higher homotopy groups

<u>3.4 Stable homotopy groups of spheres.</u> <u>There are isomorphisms</u>

$$\pi_i Z_\infty K(S_\infty, 1) \approx \pi_i (\Omega^\infty S^\infty)_0 \qquad i \geq 1$$

$$\pi_i Z_\infty K(A_\infty, 1) \approx \pi_i (\Omega^\infty S^\infty)_0 \qquad i \geq 2$$

where $(\Omega^\infty S^\infty)_0$ denotes the constant component of $\Omega^\infty S^\infty = \varinjlim \Omega^n S^n$.

The first part is a consequence of Ch.I, 5.5, Ch.V, 3.3 and the fact that [Priddy] there is a map

$$|K(S_\infty, 1)| \longrightarrow (\Omega^\infty S^\infty)_0$$

which induces an isomorphism on integral homology. The second half follows from the first by applying Ch.II, 5.1 to the fibration (see 3.3(iii))

$$K(A_\infty, 1) \longrightarrow K(S_\infty, 1) \longrightarrow K(Z_2, 1).$$

It is easy to deduce similar results for other coefficient rings R.

Next we briefly discuss Dror's observation that, for a perfect group G, the higher homotopy groups of $Z_\infty K(G, 1)$ can be interpreted as

3.5 Homotopy groups of simple acyclic spaces. Let G be a perfect group and let G' denote the extension of G

$$* \longrightarrow H_2G \longrightarrow G' \longrightarrow G \longrightarrow *$$

corresponding to $\mathrm{id} \in H^2(G; H_2(G; Z))$. Then G' is superperfect, i.e.

$$H_i(G'; Z) = 0 \qquad\qquad i = 1, 2$$

and hence [Dror (A)] there is, up to homotopy, a unique space A(G') such that

(i) A(G') is acyclic, i.e. $\tilde{H}_*(A(G'); Z) = 0$, and

(ii) $\pi_1 A(G') \approx G'$ and $\pi_1 A(G')$ acts trivially on $\pi_i A(G')$ for $i > 1$.

Moreover $A(G')$ is, up to homotopy, the "fibre" of the map $\phi\colon K(G', 1) \to Z_\infty K(G', 1)$ and from this it is not hard to deduce (using Ch.II, 2.2) that

$$\pi_2 Z_\infty K(G, 1) \approx H_2(G; Z)$$

$$\pi_i Z_\infty K(G, 1) \approx \pi_{i-1} A(G') \qquad i > 2.$$

We end this section with a result of [Sullivan, 4.28 ff] which shows that a non-nilpotent action of a nilpotent fundamental group on a higher homotopy group can create as much havoc as a non-nilpotent fundamental group (see above).

3.6 Classifying spaces for Z_p-completions of spheres. Let p be an odd prime and let $n \geq 2$ divide $p-1$. Then there exists a space $X \in \mathcal{J}_{*C}$ such that

(i) $\pi_1 X = Z_n$

(ii) $\pi_2 X = \hat{Z}_p$, the p-adic integers (Ch.VI, 4.1)

(iii) $\pi_i X = *$ for $i > 2$, and

(iv) there is a homotopy equivalence

$$\Omega R_\infty X \simeq R_\infty S^{2n-1} \qquad \text{where } R = Z_p$$

and thus

$$\pi_{2n} R_\infty X \approx \hat{Z}_p$$

$$\pi_i R_\infty X \approx \text{p-torsion of } \pi_{i-1} S^{2n-1} \qquad i \neq 2n.$$

To prove this one observes that $\underline{\underline{Z}}_p$ contains a (p-1)-st root of unity [Sullivan, 1.35 ff], i.e. an element $\xi \in \underline{\underline{Z}}_p$ such that $\xi^{p-1} = 1$ and the obvious map $\underline{\underline{Z}}_p \to Z_p$ carries ξ to a primitive (p-1)-st root of unity in Z_p and then constructs X as a space in which the action of $\pi_1 X$ on $\pi_2 X$ corresponds to that of $\{1, \xi^{(p-1)/n}, \xi^{2(p-1)/n}, \cdots\}$ on $\underline{\underline{Z}}_p$. Since $\pi_1 X$ is Z_p-perfect, it follows from 3.2 that $R_\infty X$ is simply connected and that $H^*(R_\infty X; Z_p) \approx H^*(X; Z_p)$. An easy computation shows that $H^*(X; Z_p)$ is a Z_p-polynomial algebra on a generator of degree $2n$ and hence $\Omega R_\infty X$ is a $(2n-1)$-connected Z_p-complete space and $H^*(\Omega R_\infty X; Z_p)$ is an exterior algebra on a generator in dimension $2n-1$. The desired result now follows readily.

§4. Spaces with finite homotopy or homology groups

Another class of spaces which are R-good consists of the spaces with finite homotopy groups. For such spaces the Z-completion is, up to homotopy, the product of the Z_p-completions. This last statement, in fact, holds for all spaces with finite homology groups, i.e.:

4.1 Proposition. Let J be a set of primes, let $R = Z_{(J)}$ (Ch.V, 6.1), the integers localized at J, and let $X \in \mathcal{S}_{*C}$ be such that $H_i(X; R)$ is finite for each $i \geq 1$. Then the natural map

$$R_\infty X \longrightarrow \prod_{p \in J} (Z_p)_\infty X$$

is a homotopy equivalence.

4.2 Corollary. Let J be a set of primes, let $R = Z_{(J)}$ and let $X \in \mathcal{S}_{*C}$ be such that $\pi_i X$ is finite for each $i \geq 1$. Then the natural map

$$R_\infty X \longrightarrow \prod_{p \in J} (Z_p)_\infty X$$

is a homotopy equivalence.

Proof of 4.1. By Ch.III, 6.2 and 6.5

$$(Z_p)_\infty X \simeq \varprojlim (Z_p)_\infty R_s X \qquad \text{for } p \in J$$

and, as $R_s X$ is an R-nilpotent space with finite homotopy groups,

$$R_s X \simeq \prod_{p \in J} (Z_p)_\infty R_s X \qquad \text{for } s < \infty .$$

The desired result follows easily.

Now we can state

<u>4.3 Proposition</u>. <u>Let</u> $X \in \mathscr{S}_{*c}$ <u>be such that</u> $\pi_i X$ <u>is finite for each</u> $i \geq 1$. <u>Then</u>

<u>(i)</u> $\pi_i (Z_p)_\infty X$ <u>is a finite p-group for all</u> i <u>and</u> p <u>(prime)</u>.

Hence (4.2, Ch.I, 6.1 and 7.2 and Ch.II, 5.2(iv))

<u>(ii)</u> $R_\infty X$ <u>is nilpotent for</u> $R \subset Q$ <u>and</u> $R = Z_p$,

and therefore (Ch.I, 5.2, Ch.V, 3.4 and Ch.VI, 5.3)

<u>(iii)</u> X <u>is R-good for</u> $R \subset Q$ <u>and</u> $R = Z_p$.

<u>Proof</u>. It suffices to construct, for each prime p, a map $X \to Y \in \mathscr{S}_{*c}$ which induces an isomorphism $H_*(X; Z_p) \approx H_*(Y; Z_p)$ and is such that $\pi_i Y$ is a finite p-group for all i. This can be done by "attaching Moore cells" as follows.

Let n be the smallest integer such that $\pi_n X$ is not a p-group and let $\alpha \in \pi_n X$ be an element of order k, prime to p. We may suppose X to be fibrant and choose a map

$$u: \quad M(Z_k, n) \longrightarrow X \qquad\qquad \in \mathscr{S}_{*c}$$

representing α, where $M(Z_k, n)$ is a Moore space of type (Z_k, n). If C_u denotes the mapping cone of u, then the inclusion $X \to C_u$ clearly induces isomorphisms

$$H_*(X; Z_p) \approx H_*(C_u; Z_p)$$

$$\pi_i X \approx \pi_i C_u \qquad\qquad \text{for } i < n$$

and an epimorphism $\pi_n X \to \pi_n C_u$ which annihilates $\alpha \in \pi_n X$. Moreover we will show below that $\pi_i C_u$ is finite for all i and iteration of the above construction thus yields the desired map $X \to Y$.

To show that the $\pi_i C_u$ are finite, one considers the universal covering $f: \tilde{C}_u \to C_u$ and observes that $\tilde{C}_u$ can be obtained from its subspace $f^{-1}(X)$ by "attaching Moore cells" for each of the liftings in the diagram

$$\begin{array}{ccc} & & f^{-1}(X) \\ & \nearrow & \big\downarrow f \\ M(Z_k, n) & \xrightarrow{u} & X \end{array} .$$

Since there are only finitely many such liftings and since $f^{-1}(X)$ has finite homotopy (and hence homology) groups, it is clear that $\tilde{C}_u$ has finite homology (and hence homotopy) groups. Consequently $\pi_i C_u$ is finite for all i.

As an illustration of 4.2 and 4.3 we investigate the *Z-completion* of $K(S_3, 1)$ where S_3 denotes the *symmetric group of degree 3* (see 3.3(iii)) and prove

4.4 Proposition.

$$Z_\infty K(S_3, 1) \simeq (Z_2)_\infty K(S_3, 1) \times (Z_3)_\infty K(S_3, 1)$$

where

$$(Z_2)_\infty K(S_3, 1) \simeq K(Z_2, 1)$$

and there is a fibration, up to homotopy,

$$(Z_3)_\infty S^3 \xrightarrow{j} (Z_3)_\infty S^3 \longrightarrow (Z_3)_\infty K(S_3, 1)$$

<u>in which j is of degree 3.</u>

<u>Proof</u>. The first statement follows from 4.2 and the fact that $K(S_3, 1)$ is Z_p-acyclic for $p \neq 2, 3$, while the second is true, because the obvious map $K(S_3, 1) \to K(Z_2, 1)$ is a Z_2-homology equivalence.

To get a hold on $(Z_3)_\infty K(S_3, 1)$ one applies 4.3 to the obvious (co-)homology data for $K(S_3, 1)$ and finds that $(Z_3)_\infty K(S_3, 1)$ is a Z_3-complete space with

$$\begin{aligned} \pi_i (Z_3)_\infty K(S_3, 1) &= * & \text{for } i < 3 \\ &= Z_3 & \text{for } i = 3 \end{aligned}$$

and that the algebra $H^*((Z_3)_\infty K(S_3, 1); Z_3)$ factors as a tensor product of an exterior algebra on a 3-dimensional generator with a polynomial algebra on a 4-dimensional generator. From this it is not hard to obtain the desired result.

We end with a general

<u>4.5 Remark</u>. For the spaces considered in this section one can obtain more information on $\pi_* R_\infty X$ by combining 4.1 with the homotopy spectral sequences $\{E_r(X; Z_p)\}$ (Ch.I, 4.4) as one has (Ch.VI, 9.3). <u>If</u> $X \in \mathcal{S}_{*c}$ <u>is such that</u> $H_i(X; Z_p)$ <u>is finite for each</u> $i \geq 1$, <u>then the spectral sequence</u> $\{E_r(X; Z_p)\}$ <u>is Mittag-Leffler in all dimensions</u> ≥ 1 <u>(Ch.IX, 5.5), and thus converges completely to</u> $\pi_*(Z_p)_\infty X$.

§5. Spaces with a finite fundamental group

In this section we show that spaces with a finite fundamental group are Z_p-good for all primes p. However such a space need not be Z-good; the projective plane P^2 already provides a counter example.

5.1 Proposition. Let $X \in \mathscr{S}_{*C}$ be such that $\pi_1 X$ is finite. Then X is Z_p-good for all primes p.

Proof. As in the proof of 4.3 one "attaches Moore cells" to obtain a Z_p-homology equivalence $X \to Y \in \mathscr{S}_{*C}$ such that $\pi_1 Y$ is a finite p-group; and one is thus reduced to proving that Y is Z_p-good. For this it suffices, in turn, to show that the Postnikov fibration, up to homotopy

$$\tilde{Y} \longrightarrow Y \longrightarrow K(\pi_1 Y, 1)$$

satisfies the hypotheses of Ch.II, 5.1, i.e. that $\pi_1 Y$ acts nilpotently on each $H_i(\tilde{Y}; Z_p)$. But this is indeed the case because a finite p-group G always acts nilpotently on a Z_p-module M.

To prove this last statement, observe that (Ch.II, 5.2(iv)) G acts nilpotently on the Z_p-group ring $Z_p G$. Thus, if $I \subset Z_p G$ denotes the augmentation ideal, there is an integer n such that $I^n = 0$. The desired result now follows from the fact that $I^n M \subset M$ is nothing but the n-th term in the "lower central series of M with respect to the action of G".

Unfortunately, even a finite space with a finite fundamental group need not be Z-good, as can be seen from the following:

5.2 Counter example. The projective plane P^2 is not Z-good.

5.3 Remark. If $K(F, 1)$ were Z-good for every finitely generated free group, then all spaces $X \in \mathcal{S}$ of finite type (i.e. X_n finite for all n) would also be Z-good. The above counter example thus implies that some finite wedge of circles is not Z-good.

Proof of 5.2. We want to show that $H_4(Z_\infty P^2; Q) \neq 0$. For this let $R = Z_2$. Then (4.1)

$$R_\infty P^2 \simeq Z_\infty P^2$$

and thus (Ch.II, 5.1) there is a fibration, up to homotopy

$$R_\infty S^2 \longrightarrow Z_\infty P^2 \longrightarrow K(Z_2, 1)$$

and $H_*(Z_\infty P^2; Q)$ can be identified with the quotient of $H_*(R_\infty S^2; Q)$ under the action of Z_2.

By [J.H.C. Whitehead] there is a "certain exact sequence"

$$\cdots \longrightarrow \pi_4 R_\infty S^2 \longrightarrow H_4(R_\infty S^2; Z) \longrightarrow \Gamma(\pi_2 R_\infty S^2) \longrightarrow \pi_3 R_\infty S^2 \longrightarrow \cdots$$

where Γ is the functor which assigns to an abelian group A, the abelian group $\Gamma(A)$ with a generator $\gamma(x)$ for each $x \in A$ and relations

$$\gamma(x) = \gamma(-x)$$

$$\gamma(x+y+z) - \gamma(x+y) - \gamma(y+z) - \gamma(z+x) + \gamma(x) + \gamma(y) + \gamma(z) = 0$$

for all $x, y, z \in A$. Tensoring this with Q, we obtain an exact sequence

$$0 \longrightarrow H_4(R_\infty S^2; Q) \longrightarrow Q \otimes \Gamma(\pi_2 R_\infty S^2) \longrightarrow Q \otimes \pi_3 R_\infty S^2 \longrightarrow \cdots$$

and as Z_2 acts trivially on $Q \otimes \Gamma(\pi_2 R_\infty S^2)$ it follows that

$$H_4(R_\infty S^2; Q) \approx H_4(Z_\infty P^2; Q).$$

Moreover a close inspection shows that the above map

$$Q \otimes \Gamma(\pi_2 R_\infty S^2) \longrightarrow Q \otimes \pi_3 R_\infty S^2$$

corresponds to the map $\Gamma(Q \otimes \underline{\underline{Z}}_2) \to Q \otimes \underline{\underline{Z}}_2$ which sends $\gamma(x)$ to x^2 for each $x \in Q \otimes \underline{\underline{Z}}_2$ ($Q \otimes \underline{\underline{Z}}_2$ is, of course, the field of 2-adic numbers) and it thus remains to show that the map $\Gamma(Q \otimes \underline{\underline{Z}}_2) \to Q \otimes \underline{\underline{Z}}_2$ has non-zero kernel. To do this we choose an element $a \in Q \otimes \underline{\underline{Z}}_2$ such that a, a^2 and a^3 are linearly independent over Q. This is possible, by a cardinality argument, since each equation $b_3 x^3 + b_2 x^2 + b_1 x = 0$ has only finitely many solutions $x \in Q \otimes \underline{\underline{Z}}_2$. The results of [J.H.C. Whitehead, §5] then show that

$$\gamma(a+a^3) - \gamma(a) - \gamma(a^3) - 2\gamma(a^2) \qquad \in \Gamma(Q \otimes \underline{\underline{Z}}_2)$$

is a non-zero element in the kernel of $\Gamma(Q \otimes \underline{\underline{Z}}_2) \to Q \otimes \underline{\underline{Z}}_2$.

Actually the above argument shows that <u>the projective plane P^2 is not $Z_{(J)}$-good if $2 \in J$</u>.

§6. R-homotopy theories

We end this chapter with the observation that there are such things as

6.1 R-homotopy theories for $R \subset Q$ and $R = Z_p$. By this we mean that it is possible to define in the category of spaces $\mathcal{S}$ notions of *weak R-equivalence, R-cofibration* and *R-fibration* such that:

(i) these notions satisfy Quillen's axioms for a closed simplicial model category (Ch.VIII, 3.5), and

(ii) a map between simply connected spaces is a weak R-equivalence if and only if it induces an isomorphism on R-homology.

In fact, these notions can be defined in such a manner that in addition

(iii) a map $X \to * \in \mathcal{S}$ is a weak R-equivalence if and only if X is R-acyclic, i.e. $\tilde{H}_*(X; R) = *$.

Our main tool for proving this will be

6.2 A partial R-completion functor C^R for $R \subset Q$ and $R = Z_p$. This will be a variation of the functor R_∞ in which "part of the fundamental group is not completed", with the result that the natural map $C^R X \to (C^R)^2 X$ is always a homotopy equivalence. In more detail:

Let P denote the functor which associates with every group π its *maximal R-perfect subgroup* i.e. (3.1) the largest subgroup $G \subset \pi$ for which $H_1(G; R) = 0$. (Clearly such a maximal R-perfect subgroup exists and is unique). Next, for $X \in \mathcal{S}$, let $\mathrm{Sin}|X|$ be the singular complex of its realization (Ch.VIII, §2) and let $\mathrm{Sin}|X|/P$ denote the space obtained from this by "killing, in each component, the higher homotopy groups and the maximal R-perfect subgroup of the fundamental group", i.e. by identifying two n-simplices $u, v \in \mathrm{Sin}|X|$ whenever, for every sequence of integers $(i_1, \cdots, i_{n-1})$ with $0 \le i_1 < \cdots < i_{n-1} \le n$

(i) the 1-simplices $d_{i_1} \cdots d_{i_{n-1}} u$ and $d_{i_1} \cdots d_{i_{n-1}} v$ have the same vertices, and

(ii) these two 1-simplices "differ" by an element of the maximal R-perfect subgroup of the fundamental group (of their component).

The <u>partial R-completion</u> $C^R X$ of X now is obtained by fibre-wise R-completion of the fibration $\mathrm{Sin}|X| \to \mathrm{Sin}|X|/P$, i.e. by putting (Ch.I, §8)

$$C^R X = \dot{R}_\infty \mathrm{Sin}|X|.$$

This partial R-completion comes with an obvious map (see Ch.I, §8 and Ch.VIII, §2)

$$\phi: X \longrightarrow C^R X \qquad \in \mathcal{S}$$

which has the following useful properties:

<u>6.3 Proposition</u>. <u>Let</u> $X \in \mathcal{S}_{*C}$. <u>Then the map</u> $\phi: X \to C^R X$ <u>induces isomorphisms</u>

$$\pi_1 X/P\pi_1 X \approx \pi_1 C^R X$$

$$H_*(X; R(\pi_1 X/P\pi_1 X)) \approx H_*(C^R X; R(\pi_1 C^R X)) \qquad \text{(twisted coefficients)}$$

<u>where</u> $R(-)$ <u>denotes the group ring over</u> R <u>and the twisted coefficients are the obvious ones.</u>

6.4 Proposition. For all $X \in \mathcal{S}$, the natural map $\phi: C^R X \to (C^R)^2 X$ is a weak equivalence.

Proof. These propositions follow readily from 3.2 and Ch.I, §8 and the fact that the homology with twisted coefficients $H_*(X; R(\pi_1 X/P\pi_1 X))$ is isomorphic with the ordinary homology $H_*(F; R)$ where F denotes the fibre of the fibration $\mathrm{Sin}|X| \to \mathrm{Sin}|X|/P$.

Now we are ready to define:

6.5 Weak R-equivalences, R-cofibrations and R-fibrations. A map $f: X \to Y \in \mathcal{S}$ will be called a weak R-equivalence if the induced map $C^R f: C^R X \to C^R Y \in \mathcal{S}$ is a weak equivalence. Thus, in view of 6.3 and Ch.I, 7.1, a map $f: X \to Y \in \mathcal{S}$ is a weak R-equivalence if and only if it is the disjoint union of maps $f_b: X_b \to Y_b$ between connected spaces, each of which induces an isomorphism

$$\pi_1 X_b/P\pi_1 X_b \approx \pi_1 Y_b/P\pi_1 Y_b$$

and an isomorphism of homology with twisted coefficients

$$H_*(X_b; R(\pi_1 X_b/P\pi_1 X_b)) \approx H_*(Y_b; R(\pi_1 Y_b/P\pi_1 Y_b)).$$

A map in $\mathcal{S}$ will be called an R-cofibration if it is a cofibration (i.e. injection) in $\mathcal{S}$ and a map in $\mathcal{S}$ will be called an R-fibration if it has the right lifting property with respect to all R-cofibrations which are weak R-equivalences. A simple obstruction argument then implies that every fibration $X \to Y \in \mathcal{S}_{*c}$ for which $P\pi_1 X = * = P\pi_1 Y$, is an R-fibration, and so is every pull back of such a fibration.

<u>Proof of 6.1(i), (ii) and (iii)</u>. Parts (ii) and (iii) follow from 6.5, while the axioms for a closed model category (Ch.VIII, 3.5) are easily verified, except for the second factorization axiom CM5(ii).

To deal with this consider, for a map $f: X \to Y \in \mathcal{S}$, the commutative diagram

$$\begin{array}{ccccccc} X & & \longrightarrow & & & & C^R X \\ \downarrow & & & & & & \downarrow \simeq \\ & & X''' & \longrightarrow & X'' & \longrightarrow & X' \\ \downarrow & & \downarrow & & \downarrow & & \downarrow \\ Y & \xrightarrow{=} & Y & \longrightarrow & Y' & \xrightarrow{\simeq} & C^R Y \end{array}$$

where $X \to C^R X$, $Y \to C^R Y$ and $C^R X \to C^R Y$ are the obvious maps, $C^R X \to X'$ is a cofibration and a weak equivalence, $X' \to C^R Y$ is a fibration, $Y \to Y'$ is a cofibration, $Y' \to C^R Y$ is a fibration and a weak equivalence and $X''' \to Y$ and $X'' \to Y'$ are pull backs of the map $X' \to C^R Y$. By 6.5 the map $X' \to C^R Y$ is an R-fibration and so is therefore the map $X''' \to Y$. Furthermore the fibration $Y' \to C^R Y$ induces a fibration over the universal covering of $C^R Y$ with the same fibres and it follows readily from the Serre spectral sequence for this induced fibration and the fact that the map $Y' \to C^R Y$ is a weak equivalence, that all these fibres are R-acyclic. And as these fibres are also the fibres of the fibration $X'' \to X'$ one gets, by reversing this argument, that the map $X'' \to X'$ is a weak R-equivalence. Thus the map $X \to X'''$ is a weak R-equivalence and a factorization of this map into a cofibration and a fibration which is a weak equivalence now gives rise to the desired factorization of the map $X \to Y$.

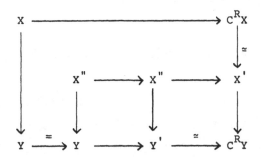

As an application we consider:

6.6 R-homotopy groups. Given the above model category for R-homotopy theory one can [Quillen (HA)], for $X \in \mathcal{S}_*$, define its *R-homotopy groups* $\pi_*^R X$ by $\pi_*^R X = \pi_* C^R X$. The following examples then show that the R-homotopy groups of a space need not coincide with its (ordinary) homotopy groups, even if $R = Z$.

(i) $\pi_1^R X \approx \pi_1 X / P\pi_1 X$.

(ii) If X is simply connected and $R \subset Q$, then (Ch.V, 3.1) $\pi_*^R X \approx R \otimes \pi_* X$.

(iii) If X is simply connected, $\pi_i X$ *is finitely generated for each i, and* $R = Z_p$, *then (Ch.VI, 5.2)* $\pi_*^R X \approx \hat{Z}_p \otimes \pi_* X$.

(iv) If S_∞ *denotes the infinite symmetric group, then (3.4)* $\pi_*^Z K(S_\infty, 1) \approx \pi_*(\Omega^\infty S^\infty)_0$.

However, just like the ordinary homotopy groups [Kan (AX)], *the R-homotopy groups can be characterized by four simple axioms*. All one has to do for this is to replace everywhere in [Kan (AX)] "fibration" by "R-fibration" and "weak homotopy equivalence" by "weak R-equivalence".

The partial R-completion is closely related to:

6.7 The R-acyclic functor. For $X \in \mathcal{S}_{*C}$, let

$$\alpha: A^R X \longrightarrow X \qquad \in \mathcal{S}_{*C}$$

be the fibration induced by the map $\phi: X \to C^R X$ from the path fibration [May, p.99] over $C^R X$. Then, as in 6.3, one readily proves:

(i) $\tilde{H}_*(A^R X; R) = *$ *for all* $X \in \mathcal{S}_{*C}$, and

(ii) the map $\alpha: A^R X \to X$ is a weak equivalence if and only if $\tilde{H}_*(X; R) = *$.

It is not hard to see that this implies that, up to homotopy, $A^R X$ is the maximal R-acyclic subspace of X, i.e. for every $B \in \mathcal{J}_{*C}$ with $\tilde{H}_*(B; R) = *$ and every map $f: B \to X \in \mathcal{J}_{*C}$ there is a unique homotopy class of maps $g: B \to A^R X$ such that $f \sim \alpha g$.

For $R = Z$ the acyclic functor is due to [Dror (A)] and was used by him to analyze the structure of acyclic spaces.

We end with a comment on possible

6.8 Variations. Except for 6.1(iii) and 6.7, the above results remain true if the functor P of 6.2 is not required to be maximal. For instance, if one takes for P the functor which assigns to every group its trivial subgroup, then, for $X \in \mathcal{J}_{*C}$, the resulting partial R-completion has the same fundamental group as X and its universal covering space [Lamotke, Ch.III] has the same homotopy type as the R-completion of the universal covering of X. The resulting Z-homotopy theory thus is nothing but the ordinary homotopy theory.

Of course, for fixed R, all the different choices of the functor P yield the same R-homotopy theory for simply connected spaces and for spectra.

Part II. Towers of fibrations, cosimplicial spaces and homotopy limits

§0. Introduction to Part II

In Part II of these notes we have assembled some results on towers of fibrations, cosimplicial spaces and homotopy limits (inverse and direct) which were needed in our discussion of completions and localizations in Part I, but which seem to be of some interest in themselves. More specifically:

Chapter VIII. Simplicial sets and topological spaces. This chapter does not really contain anything new. It is mainly intended to help make these notes accessible to a reader who knows homotopy theory, but who is not too familiar with the simplicial techniques which we use throughout these notes.

We point out that, in a certain precise sense, there is an equivalence between the homotopy theories of simplicial sets and topological spaces (or CW-complexes); and thus, for homotopy theoretic purposes, it does not really matter whether one uses simplicial sets or topological spaces. To emphasize this, we will throughout these notes (except in Chapter VIII where it might cause confusion) often use the word

space for simplicial set.

Chapter IX. Towers of fibrations. For use in Chapter X, we slightly generalize here two well-known results for a pointed tower of fibrations $\{X_n\}$:

(i) We show that the short exact sequence

$$* \longrightarrow \varprojlim{}^1 \pi_{i+1} X_n \longrightarrow \pi_i \varprojlim X_n \longrightarrow \varprojlim \pi_i X_n \longrightarrow *$$

also exists for i = 0. For this, of course, we first have to define a suitable notion of $\varprojlim^1$ for not necessarily abelian groups.

(ii) We generalize the usual homotopy spectral sequence to an "extended" homotopy spectral sequence, which in dimension 1 consists of (possibly non-abelian) groups, and in dimension 0 of pointed sets, acted on by the groups in dimension 1. This we do by carefully analyzing the low-dimensional part of the homotopy sequences of the fibrations $X_n \longrightarrow X_{n-1}$.

At the end of the chapter we show how these results can be used to get information on the homotopy type of the inverse limit space $\varprojlim X_n$.

Chapter X. Cosimplicial spaces. This chapter is concerned with our basic tool: cosimplicial (diagrams of) spaces.

In Part I of these notes (in Chapter I), we defined, for a ring R, the R-completion of a space X as the so-called "total space" of a certain cosimplicial space $\underset{\sim}{R}X$, and in order to prove some of the basic properties of this R-completion we needed, not surprisingly, various results on cosimplicial spaces. Those results are proved in this chapter. We

(i) lay the foundations for a homotopy theory of cosimplicial spaces, and

(ii) combining this with the results of Chapter IX, obtain, for every cosimplicial (pointed) space, an extended homotopy spectral sequence which in many cases (and in particular for $\underset{\sim}{R}X$)

gives useful information on the homotopy type of the total space.

Chapter XI. Homotopy inverse limits. In this chapter we extensively discuss a notion of homotopy inverse limits which gets around the difficulty that, in general, inverse limits do not exist in the homotopy category.

While this is of interest in itself, our main reasons for including a (rather long) chapter on this subject are that:

(i) homotopy inverse limits are closely related to cosimplicial spaces, and the results of this chapter put some of the results of the Chapters IX and X in perspective, and

(ii) we show in this chapter that, up to homotopy, the R-completion of a space X (which was defined in Chapter I as the total space of the cosimplicial $\underset{\sim}{R}X$), is indeed an R-completion of X, in the sense that it is a homotopy inverse limit of the "Artin-Mazur-like" diagram of "target spaces of maps from X to simplicial R-modules"; and this takes (some of) the mystery out of our definition of R-completion.

Moreover we show that:

(iii) the homotopy groups of homotopy inverse limits are quite accessible and there is an extended homotopy spectral sequence for approaching them,

(iv) homotopy inverse limits are closely related to the derived functors $\varprojlim^s$ of the inverse limit functor for abelian groups; and this can be used to extend the definition of $\varprojlim^1$ which we gave in Chapter IX for towers of not necessarily abelian groups, to arbitrary small diagrams,

(v) for a tower of fibrations, the homotopy inverse limit has the same homotopy type as the (ordinary) inverse limit space, and the spectral sequence for the homotopy inverse limit reduces to the short exact sequences of Chapter IX,

(vi) for many cosimplicial spaces (and in particular for $\underset{\sim}{R}X$) the homotopy inverse limit has the same homotopy type as the total space, and the homotopy spectral sequence for the homotopy inverse limit coincides, from E_2 on, with the spectral sequence of Chapter X, and

(vii) there is a cofinality theorem, which enables us to compare homotopy inverse limits for small diagrams of different "shapes", and which we use to show that, for certain large diagrams of spaces, one can, at least up to homotopy, talk of their homotopy inverse limits.

Chapter XII. Homotopy direct limits. Here we briefly discuss the dual notion of homotopy direct limits. We do this mainly for completeness' sake, although a few of the results of this chapter are used in Chapter XI in the proof of (ii).

In writing Part II we have been especially influenced by the work and ideas of Don Anderson and Dan Quillen.

Chapter VIII. Simplicial sets and topological spaces

§1. Introduction

The purpose of this chapter is

(i) to review some of the basic notions of simplicial homotopy theory, and

(ii) to convince (or at least try to convince) the reader that this simplicial homotopy theory is equivalent to the usual topological homotopy theory.

In slightly more detail:

§2. Here we define simplicial sets, give a few examples and construct the singular and realization functors between the category $\mathcal{S}$ of simplicial sets and the category $\mathcal{T}$ of topological spaces.

§3 contains Quillen's precise formulation of the sense in which the singular and realization functors induce an "equivalence between the homotopy theories of the categories $\mathcal{S}$ and $\mathcal{T}$". For this one needs in both categories notions of fibrations, cofibrations and weak equivalences.

§4. We end the chapter with a discussion of the homotopy relation for simplicial maps and review the related notion of function spaces for simplicial sets.

For a more detailed account of simplicial homotopy theory the reader may consult [May], [Lamotke], [Curtis (S)], [Gugenheim], [Quillen (HA)] and others.

§2. Simplicial sets

In this section we

(i) recall a definition of *simplicial sets* and, more generally, of *simplicial objects over an arbitrary category*,

(ii) discuss some simple *examples* of simplicial sets, and

(iii) observe that the categories $\mathcal{S}$ of simplicial sets and $\mathcal{T}$ of topological spaces are related by a pair of *adjoint functors*

$$\mathcal{S} \underset{\mathrm{Sin}}{\overset{|\ \ |}{\rightleftarrows}} \mathcal{T}$$

the *realization* functor $|\ \ |: \mathcal{S} \to \mathcal{T}$ and the *singular* functor $\mathrm{Sin}: \mathcal{T} \to \mathcal{S}$.

We start with

2.1 Simplicial objects and maps. A *simplicial object* X over a category $\mathcal{C}$ consists of

(i) for every integer $n \geq 0$ an object $X_n \in \mathcal{C}$, and

(ii) for every pair of integers (i,n) with $0 \leq i \leq n$, *face* and *degeneracy* maps

$$d_i: X_n \longrightarrow X_{n-1} \quad \text{and} \quad s_i: X_n \longrightarrow X_{n+1} \qquad \varepsilon$$

satisfying the *simplicial identities*:

$$
\begin{aligned}
d_i d_j &= d_{j-1} d_i && \text{for } i < j \\
d_i s_j &= s_{j-1} d_i && \text{for } i < j \\
&= \mathrm{id} && \text{for } i = j,\ j+1 \\
&= s_j d_{i-1} && \text{for } i > j+1 \\
s_i s_j &= s_j s_{i-1} && \text{for } i > j
\end{aligned}
$$

Similarly a simplicial map $f\colon X \to Y$ between two simplicial objects consists of maps

$$f\colon X_n \longrightarrow Y_n \qquad \in \mathcal{C}$$

which commute with the face and degeneracy maps, i.e.

$$d_i f = f d_i \quad \text{and} \quad s_i f = f s_i \qquad \text{for all } i.$$

We now specialize to

2.2 Simplicial sets. A simplicial object over the category of sets will be called a simplicial set, and we denote the category of simplicial sets by $\mathcal{S}$.

For $X \in \mathcal{S}$, the elements of X_n are called n-simplices; 0-simplices are sometimes called vertices.

There are two kinds of simplices:

2.3 Degenerate and non-degenerate simplices. For $X \in \mathcal{S}$, a simplex $x \in X$ is called degenerate if $x = s_i x'$ for some $x' \in X$ and i. Otherwise it is called non-degenerate.

The following property of degenerate simplices is very useful and not hard to verify.

Every degenerate $x \in X$ has a unique decomposition

$$x = s_{i_n} \dots s_{i_1} x'$$

such that $i_n > \dots > i_1$ and $x' \in X$ is non-degenerate. Moreover $i_1, \dots, i_n$ are precisely the "directions" in which x is degenerate, i.e. x is in the image of s_k if and only if $k \in \{i_1, \dots, i_n\}$.

This implies, for instance, that the product $X \times Y \in \mathcal{J}$ of two simplicial sets X and Y (which is defined by

$$(X \times Y)_n = X_n \times Y_n \qquad \text{for all } n$$

and the obvious face and degeneracy maps) can contain a non-degenerate simplex (x,y) for which both $x \in X$ and $y \in Y$ are degenerate (but in different "directions").

One can get a better idea what, in general, a simplicial set looks like by considering the singular and realization functors between the category $\mathcal{J}$ of simplicial sets and the category $\mathcal{T}$ of topological spaces. To define these we need

2.4 The topological standard simplices. For every $n \geq 0$, the topological n-simplex, $\underline{\Delta}[n]$, is the subspace of (n+1)-dimensional Euclidean space consisting of the points $(t_0, \dots, t_n)$ for which $\Sigma\, t_i = 1$ and $0 \leq t_i \leq 1$ for all i. Similarly for all $0 \leq i \leq n$, the standard maps

$$\underline{d}^i\colon \underline{\Delta}[n-1] \longrightarrow \underline{\Delta}[n] \qquad \underline{s}^i\colon \underline{\Delta}[n+1] \longrightarrow \underline{\Delta}[n]$$

are given by the formulas

$$\underline{d}^i(t_0, \dots, t_{n-1}) = (t_0, \dots, t_i, 0, t_{i+1}, \dots, t_{n-1})$$

$$\underline{s}^i(t_0, \dots, t_{n+1}) = (t_0, \dots, t_i + t_{i+1}, \dots, t_{n+1})$$

and it is easy to check that these standard maps satisfy the dual of the simplicial identities (2.1), i.e.

$$\underline{d}^j\underline{d}^i = \underline{d}^i\underline{d}^{j-1} \qquad \text{for } i < j$$

$$\underline{s}^j\underline{d}^i = \underline{d}^i\underline{s}^{j-1} \qquad \text{for } i < j$$

$$= id \qquad \text{for } i = j, j+1$$

$$= \underline{d}^{i-1}\underline{s}^j \qquad \text{for } i > j+1$$

$$\underline{s}^j\underline{s}^i = \underline{s}^{i-1}\underline{s}^j \qquad \text{for } i > j$$

2.5 The singular functor. The singular functor

$$\text{Sin}: \mathcal{T} \longrightarrow \mathcal{S}$$

is defined as follows. For $X \in \mathcal{T}$, an n-simplex of Sin X is any map

$$\underline{\Delta}[n] \xrightarrow{x} X \in \mathcal{T}$$

while its faces $d_i x$ and its degeneracies $s_i x$ are the compositions

$$\underline{\Delta}[n-1] \xrightarrow{\underline{d}^i} \underline{\Delta}[n] \xrightarrow{x} X \qquad \underline{\Delta}[n+1] \xrightarrow{\underline{s}^i} \underline{\Delta}[n] \xrightarrow{x} X$$

Similarly, for a map $f: X \to Y \in \mathcal{T}$ and an n-simplex $x \in \text{Sin } X$, the n-simplex $(\text{Sin } f)x \in \text{Sin } Y$ will be the composition

$$\underline{\Delta}[n] \xrightarrow{x} X \xrightarrow{f} Y \qquad \in \mathcal{T}$$

Closely related to the singular functor is

2.6 The realization functor. This is the functor

$$|\ |: \mathcal{S} \longrightarrow \mathcal{T}$$

defined as follows. For $X \in \mathscr{S}$, the realization $|X|$ is obtained from the disjoint union space

$$\coprod_n X_n \times \underline{\Delta}[n]$$

by taking the identification space under the relations

$$(d_i x,u) \sim (x,\underline{d}^i u) \qquad \text{for } x \in X_{n+1},\ u \in \underline{\Delta}[n]$$

$$(s_i x,u) \sim (x,\underline{s}^i u) \qquad \text{for } x \in X_{n-1},\ u \in \underline{\Delta}[n]$$

(in this construction X_n is given the discrete topology). One can show [May, p. 56]:

For every simplicial set $X \in \mathscr{S}$, its realization $|X|$ is a CW-complex with one n-cell for every non-degenerate n-simplex of X.

The functors Sin and | | determine each other because of

2.7 The adjointness of | | and Sin. The above definitions readily imply that [May, p. 61]:

The realization functor is left adjoint to the singular functor, i.e. for $X \in \mathscr{S}$ and $Y \in \mathscr{T}$ there is a natural 1-1 correspondence between the maps

$$|X| \longrightarrow Y \qquad \qquad \in \mathscr{T}$$

and

$$X \longrightarrow \operatorname{Sin} Y \qquad \qquad \in \mathscr{S}$$

Corresponding maps are called adjoint. In particular, the adjoint of a map $f\colon |X| \to Y \in \mathscr{T}$ will send $x \in X_n$ to the simplex of Sin Y given by the composition

$$\underline{\Delta}[n] \xrightarrow{(x,\ )} \coprod_n X_n \times \underline{\Delta}[n] \xrightarrow{\text{identification}} |X| \xrightarrow{f} Y$$

Of special interest are the so-called adjunction maps

$$X \longrightarrow \mathrm{Sin}|X| \qquad \text{and} \qquad |\mathrm{Sin}\ Y| \longrightarrow Y$$

which are adjoint to

$$|X| \xrightarrow{\mathrm{id}} |X| \qquad \text{and} \qquad \mathrm{Sin}\ Y \xrightarrow{\mathrm{id}} \mathrm{Sin}\ Y.$$

We now consider the most obvious example of a simplicial set (and the cause of its name):

2.8 The simplicial set of an ordered simplicial complex. Let K be an ordered simplicial complex, i.e. a simplicial complex [May, p. 2] together with an ordering of its vertices. Then K gives rise to a simplicial set ΔK with as n-simplices the (n+1)-tuples $(v_0,\ldots,v_n)$ of vertices of K for which

(i) $v_0 \leq \ldots \leq v_n$, and

(ii) the set $\{v_0,\ldots,v_n\}$ is an m-simplex of K for some $m \leq n$,

and with face and degeneracy operators given by

$$d_i(v_0,\ldots,v_n) = (v_0,\ldots,v_{i-1},v_{i+1},\ldots,v_n)$$

$$s_i(v_0,\ldots,v_n) = (v_0,\ldots,v_i,v_i,\ldots,v_n)$$

It is not hard to show that ΔK has exactly one non-degenerate simplex for every simplex of K and its realization $|\Delta K|$ is nothing but the topological space usually associated with K [Spanier, p. 111].

An important special case is the analogue of the topological n-simplex (2.4).

2.9 The standard simplices $\Delta[n]$. An extremely useful simplicial set is the standard n-simplex $\Delta[n]$, where [n] denotes the ordered simplicial complex consisting of the (ordered) set $\{0,\ldots,n\}$

and all its subsets. A q-simplex of $\Delta[n]$ thus is any $(q+1)$-tuple $(a_0,\ldots,a_q)$ of integers such that $0 \le a_0 \le \ldots \le a_q \le n$. Thus *$\Delta[n]$ has exactly one non-degenerate n-simplex*, which we will denote by i_n, *and its realization $|\Delta[n]|$ is nothing but the topological standard simplex $\underline{\Delta}[n]$*.

The usefulness of the standard simplices is due to the following [May, p. 14]:

2.10 Universal property of the standard simplices. *Let* $X \in \mathscr{S}$ *and let* $x \in X_n$. *Then there is a unique map*

$$\Delta x\colon \Delta[n] \longrightarrow X \qquad \in \mathscr{S}$$

which sends i_n *into* x.

As an easy application of this, we note that the adjunction map (2.7) $X \to \operatorname{Sin}|X| \in \mathscr{S}$ is given by $x \to |\Delta x|$.

One can also, as an easy consequence of the universal property obtain

2.11 The standard maps. *The standard maps*

$$d^j = \Delta(d_j i_n)\colon\ \Delta[n-1] \longrightarrow \Delta[n] \qquad 0 \le j \le n$$

$$s^j = \Delta(s_j i_n)\colon\ \Delta[n+1] \longrightarrow \Delta[n] \qquad 0 \le j \le n$$

satisfy the dual of the simplicial identities 2.1, i.e.

$$d^j d^i = d^i d^{j-1} \qquad \text{for } i < j$$

$$s^j d^i = d^i s^{j-1} \qquad \text{for } i < j$$

$$= id \qquad \text{for } i = j, j+1$$

$$= d^{i-1} s^j \qquad \text{for } i > j+1$$

$$s^j s^i = s^{i-1} s^j \qquad \text{for } i > j$$

We end with a less obvious example of a simplicial set:

2.12 The n-sphere S^n. This is the simplicial set with only two non-degenerate simplices: a 0-simplex x and an n-simplex y with faces:

$$d_i y = s_{n-1} \cdots s_0 x \qquad \text{for all } i$$

It can be obtained from the standard simplex $\Delta[n]$ by "collapsing" its boundary $\mathring{\Delta}[n]$, i.e. its simplicial subset generated by its (n-1)-simplices $d_0 i_n, \ldots, d_n i_n$.

Its realization $|S^n|$ is the usual CW-complex for the n-sphere consisting of a vertex and an n-cell.

We end by defining:

2.13 The n-skeleton of a simplicial set. For $X \in \mathscr{S}$, the n-skeleton $X^{[n]} \in \mathscr{S}$ is the sub-object generated by all simplices of X of dimensions $\leq n$. For example,

(i) the (n-1)-skeleton of the standard n-simplex $\Delta[n]$ is nothing but its boundary $\mathring{\Delta}[n]$ (2.12), and

(ii) for $X \in \mathscr{S}$, the realization $|X^{[n]}|$ of its n-skeleton is the n-skeleton of its realization, the CW-complex $|X|$.

§3. Equivalence of simplicial and topological homotopy theories

We recall here various results on simplicial sets and topological spaces which imply that the realization and singular functors induce an *equivalence between the homotopy theories of the categories $\mathcal{S}$ and $\mathcal{T}$* in the following sense:

(i) *Both categories are closed model categories*, i.e. in each there are notions of *weak equivalences*, *fibrations* and *cofibrations* which satisfy Quillen's axioms [Quillen (RH), p. 233] for a closed model category.

(ii) *The functors $|\;|$ and Sin both preserve weak equivalences, and both types of adjunction maps:*

$$X \longrightarrow \operatorname{Sin}|X| \in \mathcal{S} \quad \underline{\text{and}} \quad |\operatorname{Sin} Y| \longrightarrow Y \in \mathcal{T}$$

are weak equivalences.

(iii) *The functors $|\;|$ and Sin both preserve fibrations and cofibrations* (although Sin preserves cofibres only up to a weak equivalence).

According to [Quillen (HA), p. I, 1.13] (i) implies that one can (without running into set theoretical difficulties) form the *homotopy categories* $\operatorname{Ho}\mathcal{S}$ and $\operatorname{Ho}\mathcal{T}$ from $\mathcal{S}$ and $\mathcal{T}$ by *localizing with respect to* (i.e. formally inverting) *the weak equivalences*. It then follows from (ii) that *the functors $|\;|$ and Sin induce an equivalence of categories*:

$$\operatorname{Ho}\mathcal{S} \;\underset{\operatorname{Sin}}{\overset{|\;|}{\rightleftarrows}}\; \operatorname{Ho}\mathcal{T}$$

Moreover (ii) and (iii) and the adjointness of the functors $| \ |$ and Sin imply that every homotopy theoretical notion on the category $\mathcal{T}$ gives rise to a homotopically equivalent notion on the category $\mathcal{S}$ and visa versa.

We start with a brief discussion of homotopy groups, as we will use them to define weak equivalences.

3.1 Homotopy groups (and pointed sets). Although the homotopy groups of a simplicial set X can be defined "simplicially" [May, p. 7 and p. 61], it is easier to define them as the homotopy groups of the realization $|X|$. To be precise: Let $X \in \mathcal{S}$, let $* \in X$ be a base point (i.e. an arbitrary but fixed vertex) and denote also by $*$ the corresponding point $* \in |X|$. Then we put,

$$\pi_n(X,*) = \pi_n(|X|,*) \qquad \text{for all } n \geq 0$$

and, when no confusion is possible, write often,

$$\pi_n X \qquad \underline{\text{instead of}} \qquad \pi_n(X,*)$$

Now we are ready for

3.2 Weak equivalences. A map $f: X \to Y \in \mathcal{S}$ or $\mathcal{T}$ will be called a weak equivalence if f induces an isomorphism

$$\pi_n X \approx \pi_n Y$$

for every choice of base point $* \in X$ and all $n \geq 0$. Then one has [May, p. 65]:

(i) A map $f: X \to Y \in \mathcal{T}$ is a weak equivalence if and only if the map Sin f: Sin X $\to$ Sin Y $\in \mathcal{S}$ is one.

(ii) A map $f: X \to Y \in \mathcal{S}$ is a weak equivalence if and only if the map $|f|: |X| \to |Y| \in \mathcal{T}$ is one.

(iii) The adjunction maps

$$X \longrightarrow \operatorname{Sin}|X| \in \mathcal{S} \qquad \text{and} \qquad |\operatorname{Sin} Y| \longrightarrow Y \in \mathcal{T}$$

are weak equivalences for all $X \in \mathcal{S}$ and $Y \in \mathcal{T}$.

3.3 Fibrations. For $0 \leq k \leq n$ let

$$\Delta[n,k] \subset \Delta[n]$$

denote the simplicial subset generated by the simplices

$$d_0 i_n, \ldots, d_{k-1} i_n, d_{k+1} i_n, \ldots, d_n i_n$$

(i.e. $|\Delta[n,k]|$ consists of all but one face of $|\Delta[n]| = \underline{\Delta}[\Delta]$). A map $f: X \to Y \in \mathcal{S}$ then is called a *fibration* if in every (commutative) *solid* arrow diagram

$$\begin{array}{ccc} \Delta[n,k] & \longrightarrow & X \\ \downarrow & \nearrow & \downarrow \\ \Delta[n] & \longrightarrow & Y \end{array}$$

the *dotted* arrow exists. Furthermore, for every base point $* \in Y$, we will denote by the same symbol $*$ the simplicial subset of Y generated by $*$ (which consists of the simplices $s_0 \ldots s_0 *$) and call the simplicial subset $f^{-1}* \subset X$, the *fibre* of f over $*$.

These fibrations in $\mathcal{S}$ are closely related to the *(Serre) fibrations* in $\mathcal{T}$. In fact, it is clear that:

(i) A map $f: X \to Y \in \mathcal{T}$ is a fibration if and only if the map $\operatorname{Sin} f: \operatorname{Sin} X \to \operatorname{Sin} Y \in \mathcal{S}$ is one.

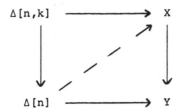

On the other hand one has [Quillen (KS)]:

(ii) If $f: X \to Y \in \mathcal{S}$ is a fibration, then so is the map $|f|: |X| \to |Y| \in \mathcal{T}$ and, for every choice of base point $* \in Y$, the inclusion $|f^{-1}*| \to |f|^{-1}*$, of "the realization of the fibre" in "the fibre of the realization", is a homeomorphism.

A convenient related notion is that of a *fibrant* object $X \in \mathcal{S}$ or $\mathcal{T}$, i.e. an object such that the (unique) map $X \to * \in \mathcal{S}$ or $\mathcal{T}$ (where $* = \Delta[0]$ or $\underline{\Delta}[0]$) is a fibration. Clearly *every topological space is fibrant*, but *not* every simplicial set, as, for instance, *$\Delta[n]$ is not fibrant for $n > 0$*. A fibrant simplicial set is also called a *Kan complex* or said to satisfy the *extension condition* [May, p. 2].

3.4 Cofibrations. A map $i: A \to B \in \mathcal{S}$ is called a *cofibration* if it is 1-1, while a map $i: A \to B \in \mathcal{T}$ will be called a *cofibration* if it has the *left lifting property* with respect to all fibrations which are weak equivalences, i.e. if for every (commutative) *solid* arrow diagram

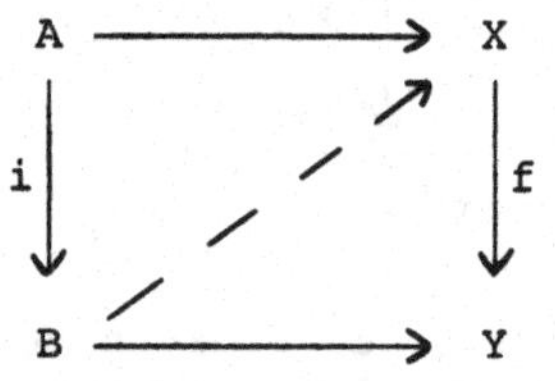

where f is a fibration which is a weak equivalence, the *dotted* arrow exists. These definitons imply:

(i) A map $i: A \to B \in \mathcal{S}$ is a cofibration if and only if the map $|i|: |A| \to |B| \in \mathcal{T}$ is one.

(ii) If $i: A \to B \in \mathcal{T}$ is a cofibration, then so is the map $\text{Sin}\, i: \text{Sin}\, A \to \text{Sin}\, B \in \mathcal{S}$, and the obvious map $\text{Sin}\, B / \text{Sin}\, A \to \text{Sin}(B/A)$

from the "cofibre of Sin i" to the "Sin of the cofibre of i" is a weak equivalence.

Again a convenient related notion is that of a cofibrant object $B \in \mathcal{S}$ or $\mathcal{T}$, i.e. an object such that the (unique) map $\phi \to B \in \mathcal{S}$ or $\mathcal{T}$ (where ϕ is empty) is a cofibration. Clearly every simplicial set is cofibrant and every CW-complex (but not every topological space) is cofibrant.

Now we can make clear what is meant by the statement that:

3.5 The categories $\mathcal{S}$ and $\mathcal{T}$ are closed model categories. According to [Quillen (HA), p. II, 3.1 and p. II, 3.14] the categories $\mathcal{S}$ and $\mathcal{T}$, with the weak equivalences, fibrations and cofibrations defined above, are closed model categories, i.e. [Quillen (RH), p. 233] they satisfy the following five axioms:

CM 1. Each category is closed under finite direct and inverse limits.

CM 2. If f and g are maps such that gf is defined, then, if two of f, g and gf are weak equivalences, so is the third.

CM 3. If f is a retract of g (i.e. if there are, in the category of maps, maps a: $f \to g$ and b: $g \to f$ such that $ba = id_f$) and g is a weak equivalence, a fibration or a cofibration, then so is f.

CM 4. (Lifting). Given a solid arrow diagram

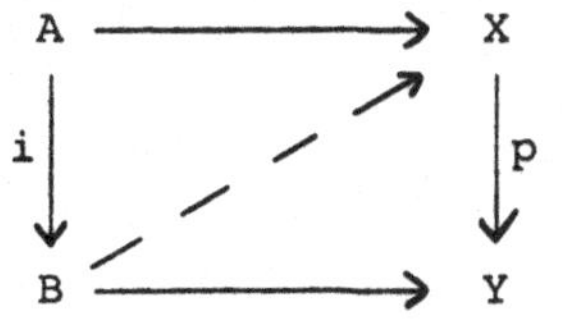

where i is a cofibration, p is a fibration, and either i or p is a weak equivalence, then the dotted arrow exists.

CM 5. (Factorization). Any map f may be factored in two ways:

(i) f = pi, where i is a cofibration and p is a fibration which is a weak equivalence.

(ii) f = pi, where p is a fibration and i is a cofibration which is a weak equivalence.

These five axioms imply [Quillen (RH), p. 234] that

(i) the class of fibrations (resp. fibrations which are weak equivalences) is closed under composition and base change and contains all isomorphisms, and dually

(ii) the class of cofibrations (resp. cofibrations which are weak equivalences) is closed under composition and co-base change and contains all isomorphisms.

Indeed, Quillen showed [Quillen (HA)] that in a closed model category one can develop much of the familiar machinery of homotopy theory, e.g. the homotopy relation for maps, loops and suspensions, fibration and cofibration exact sequences, Toda brackets, etc.

In particular we can now discuss:

3.6 The homotopy categories Ho$\mathcal{S}$ and Ho$\mathcal{T}$. These are the categories obtained from $\mathcal{S}$ and $\mathcal{T}$ by localizing with respect to (i.e. formally inverting) the weak equivalences. To be more precise we recall from [Quillen (RH), p. 208] that a localization of a category $\mathcal{C}$ with respect to a class Σ of maps in $\mathcal{C}$, consists of a category $\Sigma^{-1}\mathcal{C}$ together with a functor

$$\gamma:\mathcal{C} \longrightarrow \Sigma^{-1}\mathcal{C}$$

which carries maps in Σ into equivalences and which is universal for this property. If it exists, $\gamma:\mathcal{C} \to \Sigma^{-1}\mathcal{C}$ is an isomorphism on objects, and each map of $\Sigma^{-1}\mathcal{C}$ is a finite composition of maps of the form γg or $(\gamma u)^{-1}$ where $g \in \mathcal{C}$ and $u \in \Sigma$. We therefore can (and will) always assume that $\Sigma^{-1}\mathcal{C}$ has the same objects as $\mathcal{C}$.

In [Quillen (HA), p. I, 1.13] it is shown that any closed model category has a localization with respect to its weak equivalences; and thus the above definitions of Ho$\mathcal{S}$ and Ho$\mathcal{T}$ are legitimate. Using 3.2 it is then easy to show that the adjoint functors

$$\mathcal{S} \underset{\text{Sin}}{\overset{|\ |}{\rightleftarrows}} \mathcal{T}$$

induce an equivalence of categories

$$\text{Ho}\mathcal{S} \underset{\text{Sin}}{\overset{|\ |}{\rightleftarrows}} \text{Ho}\mathcal{T}$$

In fact, as noted at the beginning of this section, the functors | | and Sin induce an equivalence of the simplicial and topological "homotopy theories".

§4. The homotopy relation and function spaces

In the preceding section we have used weak equivalences, rather than a homotopy relation on maps, to define the homotopy categories $\mathrm{Ho}\mathcal{S}$ and $\mathrm{Ho}\mathcal{T}$; and we have thereby emphasized the underlying similarity of the simplicial and topological approaches. In this section we shall discuss the homotopy relation and show that $\mathrm{Ho}\mathcal{S}$ and $\mathrm{Ho}\mathcal{T}$ are equivalent to the "usual" homotopy categories of fibrant simplicial sets and CW-complexes. In addition we review the related topic of function spaces for simplicial sets.

We begin by disposing of the easy topological case.

4.1 The homotopy category $\mathrm{Ho}\mathcal{T}$ is equivalent to the usual CW-homotopy category, i.e. the category with CW-complexes as objects and homotopy classes of maps as maps. Moreover, for any CW-complex K and topological space X

$$\mathrm{Hom}_{\mathrm{Ho}\mathcal{T}}(K,X) \approx \{\text{homotopy classes of maps } K \longrightarrow X\}.$$

Proof. This is straightforward, using the following familiar facts:

(i) A map $X \to Y \in \mathcal{T}$ is a weak equivalence if and only if, for every CW-complex K, it induces an isomorphism between the homotopy classes of maps

$$K \longrightarrow X \quad \text{and} \quad K \longrightarrow Y$$

(ii) For every $X \in \mathcal{T}$, there is a weak equivalence $K \to X \in \mathcal{T}$, in which K is a CW-complex.

(iii) If $T: \mathcal{T} \to \mathcal{C}$ is a functor which carries weak equivalences to isomorphisms, then T carries homotopic maps to the same map.

4.2 The pointed case. In a similar way one can show that the pointed homotopy category $\mathrm{Ho}\mathcal{T}_*$ (obtained by localizing the category $\mathcal{T}_*$ of pointed topological spaces with respect to weak equivalences) is equivalent to the usual pointed CW-homotopy category.

To obtain similar results for simplicial sets we need:

4.3 The simplicial homotopy relation. Two maps

$$f_0, f_1 : X \longrightarrow Y \quad \in \mathcal{S}$$

are called homotopic if there exists a map (homotopy)

$$f : \Delta[1] \times X \longrightarrow Y \quad \in \mathcal{S}$$

which maps the "top" and "bottom" of $\Delta[1] \times X$ by f_0 and f_1 respectively, i.e. the compositions

$$X = \Delta[0] \times X \xrightarrow{d^0 \times X} \Delta[1] \times X \xrightarrow{f} Y$$

$$X = \Delta[0] \times X \xrightarrow{d^1 \times X} \Delta[1] \times X \xrightarrow{f} Y$$

are respectively equal to f_0 and f_1. When Y is fibrant, this homotopy relation is an equivalence relation and the homotopy classes of maps $X \to Y \in \mathcal{S}$ correspond to the homotopy classes of maps $|X| \to |Y| \in \mathcal{T}$.

Now we can give the simplicial analogue of 4.1

4.4 The homotopy category $\mathrm{Ho}\mathcal{S}$ is equivalent to the "usual" homotopy category of fibrant simplicial sets, i.e. the category with fibrant simplicial sets as objects and homotopy classes of maps as maps. Moreover, for X, Y $\in \mathcal{S}$ and Y fibrant

$$\mathrm{Hom}_{\mathrm{Ho}\mathcal{S}}(X,Y) \approx \{\text{homotopy classes of maps } X \longrightarrow Y\}$$

An easy consequence of this is that there are

4.5 Weak equivalences in $\mathcal{S}$ which are homotopy equivalences. If $f: X \to Y \in \mathcal{S}$ is a weak equivalence and X and Y are fibrant, then f is actually a homotopy equivalence, i.e. there is a map $g: Y \to X \in \mathcal{S}$ such that gf and fg are homotopic to the identity maps of X and Y.

4.6 The pointed case. Let $\mathcal{S}_*$ denote the category of pointed simplicial sets (= simplicial sets with base point = simplicial pointed sets). Two maps

$$f_0, f_1: X \longrightarrow Y \quad \in \mathcal{S}_*$$

then are called homotopic if there is a map (homotopy)

$$f: (\Delta[1] \times X)/(\Delta[1] \times *) \longrightarrow Y \quad \in \mathcal{S}_*$$

which maps the "top" and "bottom" of $(\Delta[1] \times X)/(\Delta[1] \times *)$ by f_0 and f_1 respectively. Again, when Y is fibrant, this is an equivalence relation, and the homotopy classes of maps $X \to Y \in \mathcal{S}_*$ correspond to the pointed homotopy classes of maps $|X| \to |Y| \in \mathcal{T}_*$. Moreover the pointed homotopy category $\text{Ho}\mathcal{S}_*$ (obtained by localizing $\mathcal{S}_*$ with respect to the weak equivalences) is equivalent to the "usual" homotopy category of pointed fibrant simplicial sets. Also, of course, $\text{Ho}\mathcal{S}_*$ is equivalent to $\text{Ho}\mathcal{T}_*$.

We conclude by reviewing the related topic of

4.7 Simplicial function spaces. For $X, Y \in \mathcal{S}$, the function space

$$\text{hom}\,(X,Y) \in \mathcal{S}$$

is the simplicial set of which an n-simplex is a map

$$\Delta[n] \times X \longrightarrow Y \quad \in \mathcal{S}$$

with as faces and degeneracies the compositions

$$\Delta[n-1] \times X \xrightarrow{d^i \times X} \Delta[n] \times X \longrightarrow Y$$

$$\Delta[n+1] \times X \xrightarrow{s^i \times X} \Delta[n] \times X \longrightarrow Y$$

Some useful properties of the function space are:

<u>(i) If Y is fibrant, then the elements of</u> $\pi_0 \hom(X,Y)$ <u>correspond to the homotopy classes of maps</u> $X \to Y \in \mathcal{S}$.

<u>(ii) If</u> $i: K \to L \in \mathcal{S}$ <u>is a cofibration and</u> $p: X \to Y \in \mathcal{S}$ <u>is a fibration, then the map</u>

$$(i,p): \hom(L,X) \longrightarrow \hom(K,X) \times_{\hom(K,Y)} \hom(L,Y) \quad \in \mathcal{S}$$

<u>is a fibration, which is a weak equivalence if either i or p is a weak equivalence.</u>

<u>(iii) For</u> $K, X, Y \in \mathcal{S}$<u>, there is a natural isomorphism</u>

$$\hom(K \times X,Y) \simeq \hom(K, \hom(X,Y)) \in \mathcal{S}$$

Similarly there are

<u>4.8 Pointed simplicial function spaces.</u> For $X, Y \in \mathcal{S}_*$, the <u>pointed function space</u>

$$\hom_*(X,Y) \quad \in \mathcal{S}_*$$

is the pointed simplicial set of which an n-simplex is a map

$$(\Delta[n] \times X)/(\Delta[n] \times *) \longrightarrow Y \quad \in \mathcal{S}_*$$

and of which the face and degeneracy maps are induced, as in 4.7, by the standard maps between the standard simplices.

Again, some useful properties are:

(i) If Y is fibrant, then the elements of $\pi_n \hom_*(X,Y)$ *correspond to the pointed homotopy classes of maps* $S^n X \to Y$, where $S^n X$ is the n-fold *reduced suspension* of X [May, p. 124].

(ii) If $i: K \to L \in \mathcal{S}_*$ *is a cofibration and* $p: X \to Y \in \mathcal{S}_*$ *is a fibration, then the map*

$$(i,p): \hom_*(L,X) \longrightarrow \hom_*(K,X) \times_{\hom_*(K,Y)} \hom_*(L,Y) \quad \in \mathcal{S}_*$$

is a fibration, which is a weak equivalence if either i or p is a weak equivalence.

(iii) For $K, X, Y \in \mathcal{S}_*$, *there is a natural isomorphism*

$$\hom_*(K \wedge X,\ Y) \approx \hom_*(K,\ \hom_*(X,Y)) \quad \in \mathcal{S}_*$$

where $K \wedge Y \in \mathcal{S}$ is the *smash product*

$$K \wedge Y = (K \times Y)/((* \times Y) \cup (K \times *))$$

4.9 Remark. The categories $\mathcal{S}$ and $\mathcal{S}_*$ are *closed simplicial model categories* in the sense of [Quillen (HA), p. II, 2.2 and (RH), p. 233], i.e. they are closed model categories with "compatible function spaces".

Chapter IX. Towers of fibrations

§1. Introduction

In this chapter we generalize two well-known results about towers of fibrations:

(i) We will show that, for a (pointed) tower of fibrations $\{X_n\}$, the short exact sequence

$$* \longrightarrow \varprojlim{}^1 \pi_{i+1} X_n \longrightarrow \pi_i \varprojlim X_n \longrightarrow \varprojlim \pi_i X_n \longrightarrow *$$

which is "well known" for $i \geq 1$, also exists for $i = 0$, if one uses a suitable notion of $\varprojlim{}^1$ for not necessarily abelian groups.

(ii) We will generalize the usual homotopy spectral sequence of a (pointed) tower of fibrations, to an "extended" homotopy spectral sequence, which in dimension 1 consists of (possibly non-abelian) groups, and in dimension 0 of pointed sets, acted on by the groups is dimension 1.

The chapter is organized as follows:

§2 and §3 deal with the first result. In §2 we discuss the functors $\varprojlim$ and $\varprojlim{}^1$ for (not necessarily abelian) groups, while §3 contains the short exact sequences and some applications.

§4 contains the construction of the extended homotopy spectral sequence.

§5 Here we show how the results of §3 and §4 can be used to get some information on the homotopy type of the inverse limit space of a tower of fibrations.

Notation and terminology. We remind the reader that these notes are written simplicially, i.e.

space = simplicial set

In particular, in this chapter, we will mainly work in the category $\mathcal{S}_*$ of pointed spaces (i.e. simplicial sets with base point), and base point preserving maps.

§2. The functors $\varprojlim$ and $\varprojlim^1$ for groups

In preparation for the decomposition of the homotopy groups of the inverse limit of a tower of fibrations into a $\varprojlim$-part and a $\varprojlim^1$-part, we discuss here in some detail:

2.1 The functors $\varprojlim$ and $\varprojlim^1$ for (not necessarily abelian) groups. A tower of (possibly non-abelian) groups and homomorphisms

$$\cdots \longrightarrow G_n \xrightarrow{j} G_{n-1} \longrightarrow \cdots \longrightarrow G_{-1} = *$$

gives rise to a left action of the product group $\prod G_n$ on the product set $\prod G_n$ given by

$$(g_0,\ldots,g_i,\ldots)\bullet(x_0,\ldots,x_i,\ldots) = (g_0x_0(jg_1)^{-1},\ldots,g_ix_i(jg_{i+1})^{-1},\ldots).$$

Clearly

$$\varprojlim G_n = \{g \in \prod G_n \mid g \circ * = *\}$$

and we define $\varprojlim^1 G_n$ as the orbit set

$$\varprojlim^1 G_n = \prod G_n \,/\, \text{action}$$

i.e. $\varprojlim^1 G_n$ is the set of equivalence classes of $\prod G_n$ under the equivalence relation given by

$$x \sim y \iff y = g \circ x \quad \text{for some } g \in \prod G_n.$$

In general $\varprojlim^1 G_n$ is only a pointed set, but if the G_n are abelian, then $\varprojlim^1 G_n$ inherits the usual (see [Milnor] and [Quillen (RH), p. 217]) abelian group structure.

It is also not hard to verify that the functors $\varprojlim$ and $\varprojlim^1$ have the following properties which are "well known" in the abelian case.

2.2 Proposition. *Let* $\{G_n\}$ *be a tower of groups, let* $k \geq 1$ *and let* $\{G_n^{(k)}\}$ *be the "k-th derived tower", i.e.*

$$G_n^{(k)} = \underline{\text{image}}\ (G_{n+k} \longrightarrow G_n).$$

Then the inclusions $G_n^{(k)} \subset G_n$ *induce isomorphisms*

$$\varprojlim G_n^{(k)} \approx \varprojlim G_n \qquad\qquad \varprojlim^1 G_n^{(k)} \approx \varprojlim^1 G_n.$$

2.3 Propositions. *A short exact sequence of towers of groups*

$$* \longrightarrow \{G_n'\} \longrightarrow \{G_n\} \longrightarrow \{G_n''\} \longrightarrow *$$

gives rise to a natural sequence of groups and pointed sets

$$* \longrightarrow \varprojlim G_n' \longrightarrow \varprojlim G_n \longrightarrow \varprojlim G_n'' \longrightarrow \varprojlim^1 G_n' \longrightarrow \varprojlim^1 G_n \longrightarrow$$

$$\varprojlim^1 G_n'' \longrightarrow *$$

which is exact in the sense that

(i) "kernel = image" at all six positions, and

(ii) the map $\varprojlim G_n'' \to \varprojlim^1 G_n'$ *extends to a natural action of* $\varprojlim G_n''$ *on* $\varprojlim^1 G_n'$ *such that elements of* $\varprojlim^1 G_n'$ *are in the same orbit if and only if they have the same image in* $\varprojlim^1 G_n$.

2.4 Proposition. *Let* $\{G_n\}$ *be a tower of groups such that each* $j: G_n \to G_{n-1}$ *is onto. Then* $\varprojlim^1 G_n = *$.

<u>2.5 Example</u>. Let Z denote the additive group of the integers and let $p^n Z \subset Z$ denote the subgroup generated by p^n. Then applying 2.3 and 2.4 to the short exact sequence of towers

$$0 \longrightarrow \{p^n Z\} \longrightarrow \{Z\} \longrightarrow \{Z/p^n Z\} \longrightarrow 0$$

for p prime, one gets

$$\varprojlim{}^1 p^n Z \approx (\varprojlim Z/p^n Z)/Z$$

$$\approx (\underline{\text{the p-adic integers}})/Z.$$

Thus $\varprojlim^1 p^n Z$ <u>is not countable</u>.

<u>2.6 Remark</u>. The towers of abelian groups form an abelian category with enough injectives; and for such towers $\varprojlim^1$ can be interpreted as the <u>first right derived functor</u> of $\varprojlim$. This follows easily using 2.3 and 2.4, since each injective is a tower of epimorphisms. (see also Ch. XI, §6.)

Also in Ch. XI, §6 we will show how to define $\varprojlim^1$ for arbitrary small diagrams of groups.

§3. The homotopy groups of the inverse limit of a tower of fibrations

We now decompose the homotopy groups of the inverse limit of a tower of fibrations into a $\varprojlim$-part and a $\varprojlim^1$-part. Various cases and applications of this have been treated by [Milnor], [Gray], [Quillen (RH), p. 217] and [Cohen].

<u>3.1 Theorem</u>. *Let* $X = \varprojlim X_n$, *where*

$$\cdots \longrightarrow X_n \xrightarrow{p} X_{n-1} \longrightarrow \cdots \longrightarrow X_{-1} = *$$

is a tower of fibrations in $\mathcal{S}_*$ *, i.e. a tower of fibrations of simplicial sets with compatible base points* $* \in X_n$. *Then there is, for every* $i \geq 0$, *a natural short exact sequence*

$$* \longrightarrow \varprojlim{}^1 \pi_{i+1} X_n \longrightarrow \pi_i X \longrightarrow \varprojlim \pi_i X_n \longrightarrow *$$

<u>3.2 Corollary</u>. *For every* $K \in \mathcal{S}_*$ *there is a natural (in the obvious sense) exact sequence of pointed sets*

$$* \longrightarrow \varprojlim{}^1 [SK, X_n] \longrightarrow [K, X] \longrightarrow \varprojlim [K, X_n] \longrightarrow *$$

where SK denotes the *reduced suspension* of K [May, p. 124] and, for Y fibrant, $[L,Y]$ stands for the *pointed set of homotopy classes* of maps $L \to Y \in \mathcal{S}_*$ (see Ch. VIII, 4.6).

This follows immediately from the fact that (see Ch. VIII, 4.8), *for* Y *fibrant, there are natural isomorphisms*

$$[K,Y] \approx \pi_0 \hom_*(K,Y) \qquad [SK,Y] \approx \pi_1 \hom_*(K,Y).$$

<u>3.3 Corollary</u>. Let $X \in \mathcal{S}_*$ <u>be fibrant, let</u>

$$K_0 \subset K_1 \subset \ldots \subset K_n \subset \ldots$$

<u>be a sequence of inclusions in</u> $\mathcal{S}_*$ <u>and let</u> $K = \varinjlim K_n$. <u>Then there is a natural exact sequence of pointed sets</u>

$$* \longrightarrow \varprojlim{}^1 [SK_n, X] \longrightarrow [K,X] \longrightarrow \varprojlim [K_n, X] \longrightarrow * .$$

Using this [Gray] shows the amusing result that there is an essential map $f\colon CP^\infty \to S^3$ such that the restrictions $f|CP^n$ are null-homotopic for all n.

<u>Proof of theorem 3.1</u>. It is easy to show that the obvious map $f\colon \pi_i X \to \varprojlim \pi_i X_n$ is onto, and it thus suffices to construct a natural isomorphism

$$g\colon \ker f \approx \varprojlim{}^1 \pi_{i+1} X_n$$

To do this, we recall [May, p. 7], that the elements of $\pi_i X$ can be considered as certain classes of i-simplices of X. Let $a \in X$ be an i-simplex representing an element $[a] \in \ker f \subset \pi_i X$ and, for each n, let a_n be its image in X_n. Then $[a_n] = * \in \pi_i X_n$ and hence one can choose a null-homotopy for a_n, i.e. an (i+1)-simplex $b_n \in X_n$ such that $d_0 b_n = a_n$ and $d_j b_n = *$ for $j > 0$. But as the (i+1)-simplices b_n and pb_{n+1} have the same faces, they determine an element of $\pi_{i+1} X_n$ which can, for instance, be obtained by choosing an (i+2)-simplex $c_n \in X_n$ such that $d_0 c_n = pb_{n+1}$, $d_1 c_n = b_n$ and $d_k c_n = *$ for $k > 2$, and then taking $[d_2 c_n] \in \pi_{i+1} X_n$. Finally we define

$$g[a] \in \varprojlim{}^1 \pi_{i+1} X_n = \prod \pi_{i+1} X_n \, / \, \underline{\text{action}}$$

as the element represented by $([d_2c_0],\ldots,[d_2c_n],\ldots)$ and a long but straightforward computation now shows that g is well-defined and has all the desired properties.

For future reference we give the following group theoretical application of theorem 3.1.

3.4 Proposition. *Let* $\{G_n\}$ *be a tower of groups. Then there is a natural isomorphism*

$$\varprojlim_n \varprojlim_k G_n^{(k)} \approx \varprojlim G_n$$

and a natural short exact sequence

$$* \longrightarrow \varprojlim{}^1_n \varprojlim_k G_n^{(k)} \longrightarrow \varprojlim{}^1 G_n \longrightarrow \varprojlim_n \varprojlim{}^1_k G_n^{(k)} \longrightarrow * \, .$$

3.5 Corollary. *If* $\{G_n\}$ *is Mittag-Leffler, i.e. if for each* n *there is an* $N < \infty$ *such that* $G_n^{(N)} = \varprojlim_k G_n^{(k)}$*, then* $\varprojlim{}^1 G_n = *$. *In particular, if each* G_n *is finite, then* $\varprojlim{}^1 G_n = *$.

Proof of 3.4. Construct a commutative lattice

$$\begin{array}{ccccccc}
 & & \downarrow & & \downarrow & & \\
\longrightarrow & & D_{k+1,n+1} & \longrightarrow & D_{k,n+1} & \longrightarrow & \\
 & & \downarrow & & \downarrow & & \\
\longrightarrow & & D_{k+1,n} & \longrightarrow & D_{k,n} & \longrightarrow & \\
 & & \downarrow & & \downarrow & &
\end{array}$$

of spaces and fibrations in $\mathcal{S}_*$ such that

(i) $D_{k,n} = *$ unless $k, n \geq 0$

(ii) for each k and n the map

$$D_{k+1,n+1} \longrightarrow D_{k+1,n} \times_{D_{k,n}} D_{k,n+1} \qquad \in \quad \mathcal{S}_*$$

is a fibration, and

(iii) $\pi_i D_{k,n} = G_n^{(k)}$ for $k, n \geq 0$, $i = 1$
$= *$ otherwise

The conditions (i) and (ii) ensure that $\{\varprojlim_k D_{k,n}\}$ and $\{\varprojlim_n D_{k,n}\}$ are towers of fibrations in $\mathcal{S}_*$. The proposition now follows from 3.1 and the fact that

$$\varprojlim_k \varprojlim_n D_{k,n} = \varprojlim_n \varprojlim_k D_{k,n} .$$

§4. The extended homotopy spectral sequence of a tower of fibrations

In this section we generalize the usual homotopy spectral sequence of a tower of fibrations to an "extended" homotopy spectral sequence, which in dimension 1 consists of (possibly non-abelian) groups, and in dimension 0 of pointed sets, acted on by the groups in dimension 1.

We start with

4.1 An observation about the homotopy sequences of a tower of fibrations. Let $\{X_n\}$ be a tower of fibrations in $\mathcal{S}_*$, i.e. a tower

$$\cdots \longrightarrow X_n \longrightarrow X_{n-1} \longrightarrow \cdots \longrightarrow X_{-1} = *$$

of fibrations with compatible base points $* \in X_n$, and let $F_n \subset X_n$ be the fibre over $*$ of the fibration $X_n \to X_{n-1}$. Then one can form the homotopy sequences [May, p. 27]

$$\cdots \longrightarrow \pi_2 X_{n-1} \longrightarrow \pi_1 F_n \longrightarrow \pi_1 X_n \longrightarrow \pi_1 X_{n-1} \longrightarrow \pi_0 F_n \longrightarrow \pi_0 X_n \longrightarrow \pi_0 X_{n-1}$$

and these sequences are "well known" to be exact in the sense that

(i) the last three objects are pointed sets, all the others are groups, and the image of $\pi_2 X_{n-1}$ lies in the center of $\pi_1 F_n$,

(ii) everywhere "kernel = image", and

(iii) the sequences come with a natural action of $\pi_1 X_{n-1}$ on $\pi_0 F_n$ which "extends" the map $\pi_1 X_{n-1} \to \pi_0 F_n$, and is such that "elements of $\pi_0 F_n$ are in the same orbit if and only if they have the same image in $\pi_0 X_n$".

From this it readily follows that one can form the r-th derived homotopy sequences $(r \geq 0)$

$$\cdots \longrightarrow \pi_2 X_{n-2r-1}^{(r)} \longrightarrow \pi_1 F_{n-r}^{(r)} \longrightarrow \pi_1 X_{n-r}^{(r)} \longrightarrow \pi_1 X_{n-r-1}^{(r)} \longrightarrow \pi_0 F_n^{(r)}$$
$$\longrightarrow \pi_0 X_n^{(r)} \longrightarrow \pi_0 X_{n-1}^{(r)}$$

where

$$\pi_i X_n^{(r)} = \underline{\text{im}}(\pi_i X_{n+r} \longrightarrow \pi_i X_n) \subset \pi_i X_n$$

$$\pi_i F_n^{(r)} = \underline{\text{ker}}(\pi_i F_n \longrightarrow \pi_i X_n / \pi_i X_n^{(r)}) / \underline{\text{action of ker}}(\pi_{i+1} X_{n-1} \longrightarrow \pi_{i+1} X_{n-r-1})$$

(for $i > 0$ the group $\pi_i F_n^{(r)}$ is the cokernel of the boundary homomorphism between the indicated kernels).

It is not hard to see that these derived homotopy sequences are also exact in the above sense. Hence one can form

4.2 The (extended) homotopy spectral sequence. For a tower of fibrations in $\mathscr{S}_*$ we define its (extended) homotopy spectral sequence $\{E_r^{s,t}\{X_n\}\}$ by

$$E_r^{s,t} = \pi_{t-s} F_s^{(r-1)} \qquad \text{for } t \geq s \geq 0, \quad r \geq 1$$

with as differentials

$$d_r\colon E_r^{s,t} \longrightarrow E_r^{s+r,t+r-1}$$

the composite maps

$$\pi_{t-s} F_s^{(r-1)} \longrightarrow \pi_{t-s} X_s^{(r-1)} \longrightarrow \pi_{t-s-1} F_{s+r}^{(r-1)} .$$

It clearly has the properties

<u>(i)</u> $E_r^{s,t}$ <u>is a group of</u> $t-s \geq 1$, <u>which is abelian if</u> $t-s \geq 2$,

<u>(ii)</u> $E_r^{s,t}$ <u>is a pointed set if</u> $t-s = 0$

<u>(iii) the differential</u> $d_r\colon E_r^{s,t} \to E_r^{s+r,t+r-1}$ <u>is a homomorphism if</u> $t-s \geq 2$, <u>and its image is a subgroup of the center of</u> $E_r^{s+r,t+r-1}$ <u>if</u> $t-s = 2$; <u>moreover</u>

$$E_{r+1}^{s,t} = (E_r^{s,t} \cap \underline{\ker}\, d_r)/(E_r^{s,t} \cap \underline{\mathrm{im}}\, d_r) \qquad t-s \geq 1$$

<u>(iv) the differential</u> $d_r\colon E_r^{s-r,s-r-1} \to E_r^{s,s}$ <u>extends to an action of</u> $E_r^{s-r,s-r+1}$ <u>on</u> $E_r^{s,s}$ <u>such that</u>

$$E_{r+1}^{s,s} \subset E_r^{s,s} \,/\, \underline{\text{action of}}\ E_r^{s-r,s-r+1} .$$

§5. Applications

The results of §3 and §4 can be used to obtain information on the homotopy type of the inverse limit space. For instance one has

5.1 *Connectivity lemma. Let* $k \geq 0$ *and* $r \geq 1$ *and let* $\{X_n\} \in \mathcal{S}_*$ *be a tower of fibrations such that* $E_r^{s,t} = *$ *for* $0 \leq t-s \leq k$. *Then*

$$\varprojlim \pi_i X_n \quad = \quad * \quad = \quad \varprojlim{}^1 \; \pi_{i+1} X_n \qquad \text{for } 0 \leq i \leq k$$

and hence (3.1) $\varprojlim X_n$ *is k-connected.*

Proof. The hypotheses imply that $\pi_i X_n^{(r-1)} = *$ and that $\pi_{i+1} X_n^{(r-1)} \to \pi_{i+1} X_{n-1}^{(r-1)}$ is onto for $0 \leq i \leq k$. The lemma then follows from 2.2 and 2.4.

5.2 *Mapping lemma. Let* $r \geq 1$, *let* $\{X_n\} \in \mathcal{S}_*$ *be a tower of fibrations such that* $E_r^{s,t} = *$ *for* $t-s = 0$ *and let* $f\colon \{X_n\} \to \{Y_n\} \in \mathcal{S}_*$ *be a map between towers of fibrations, which induces an isomorphism of the* $E_r^{s,t}$ *for all* $t-s \geq 0$. *Then* f *induces isomorphisms*

$$\varprojlim \pi_* X_n \;\approx\; \varprojlim \pi_* Y_n \qquad \varprojlim{}^1 \pi_* X_n \;\approx\; \varprojlim{}^1 \pi_* Y_n$$

and hence (3.1 and 5.1) f *induces a homotopy equivalence*

$$\varprojlim X_n \;\simeq\; \varprojlim Y_n \, .$$

Proof. The hypotheses imply that $\pi_0 X_n^{(r-1)} = * = \pi_0 Y_n^{(r-1)}$ and that $\pi_i X_n^{(r-1)} \approx \pi_i Y_n^{(r-1)}$ for $i \geq 1$ and the lemma again follows from the results of §2.

We end with a brief discussion of convergence of the spectral sequence and consider the notion of

5.3 Complete convergence. Let $\{X_n\} \in \mathcal{J}_*$ be a tower of fibrations, let $X = \varprojlim X_n$ and let

$$E_\infty^{s,t} = \varprojlim_r E_r^{s,t} = \bigcap_{r>s} E_r^{s,t} .$$

Then we will say that $\{E_r\}$ converges completely to $\pi_i X$ if, roughly speaking, $\pi_i X$ is the inverse limit of a tower of epimorphisms with the $E_\infty^{s,s+i}$ as kernels. To be more precise, form the filtration quotients

$$Q_s \pi_i X = \operatorname{im} (\pi_i X \longrightarrow \pi_i X_s)$$

and the small E_∞-terms

$$e_\infty^{s,s+i} = \ker (Q_s \pi_i X \longrightarrow Q_{s-1} \pi_i X)$$

and observe that the inclusions $Q_s \pi_i X \subset \varprojlim_r \pi_i X_s^{(r)}$ induce isomorphisms

$$\varprojlim_s Q_s \pi_i X \approx \varprojlim_s \pi_i X_s$$

and inclusions

$$e_\infty^{s,s+i} \subset E_\infty^{s,s+i} \qquad s \geq 0 .$$

We then say that $\{E_r\}$ converges completely to $\pi_i X$ $(i \geq 1)$ if

(i) $\varprojlim^1 \pi_{i+1} X_n = *$ (and hence (3.1) $\pi_i X \approx \varprojlim_s Q_s \pi_i X$)

(ii) $e_\infty^{s,s+i} \approx E_\infty^{s,s+i}$ for all $s \geq 0$.

A useful convergence test is provided by the following lemma (c.f. [Adams (AT)]).

5.4 Complete convergence lemma. *Let $\{X_n\} \in \mathcal{J}_*$ be a tower of fibrations and let $i \geq 1$. Then the condition*

$$\varprojlim_r^1 E_r^{s,s+i} = * \qquad \text{for all } s \geq 0$$

is equivalent to the combined conditions

$$\varprojlim_n^1 \pi_i X_n = *$$

$$E_\infty^{s,s+i} \approx e_\infty^{s,s+i} \qquad \text{for all } s \geq 0 .$$

In particular, if

$$\varprojlim_r^1 E_r^{s,s+i} = * = \varprojlim_r^1 E_r^{s,s+i+1} \qquad \text{for all } s \geq 0$$

then $\{E_r\}$ converges completely to $\pi_i X$.

Proof. To prove this, one combines 3.4 and the results of §2 with the existence of the short exact sequences

$$* \longrightarrow E_r^{s,s+1} \longrightarrow \pi_i X_s^{(r-1)} \longrightarrow \pi_i X_{s-1}^{(r)} \longrightarrow * \qquad (r > s).$$

Finally we observe that it is sometimes (see Ch.VI, §9) convenient to consider the slightly *stronger* notion of

5.5 Mittag-Leffler convergence. Let $\{X_n\} \in \mathcal{J}_*$ be a tower of fibrations, let $X = \varprojlim X_n$ and let

$$E_\infty^{s,t} = \varprojlim_r E_r^{s,t} = \bigcap_{r>s} E_r^{s,t} .$$

Then we say that $\{E_r\}$ <u>is Mittag-Leffler in dimension i</u> $(i \geq 1)$ if the towers $\{E_r^{s,s+i}\}_{r>s}$ are Mittag-Leffler, i.e. (3.5) if for each $s \geq 0$

$$E_{N(s)}^{s,s+i} = E_\infty^{s,s+i} \qquad \text{for some} \quad s < N(s) < \infty .$$

This definition clearly implies

<u>5.6 Mittag-Leffler convergence lemma.</u> $\{E_r\}$ <u>is Mittag-Leffler in dimension i if and only if the tower of groups</u> $\{\pi_i X_n\}$ <u>is Mittag-Leffler (3.5).</u>

<u>5.7 Proposition.</u>

<u>(i) If</u> $\{E_r\}$ <u>is Mittag-Leffler in dimension i, then</u> $\varprojlim_r^1 E_r^{s,s+i} = *$ <u>for</u> $s \geq 0$ <u>and thus (5.4)</u>

<u>(ii) if</u> $\{E_r\}$ <u>is Mittag-Leffler in dimensions i and i+1, then</u> $\{E_r\}$ <u>converges completely to</u> $\pi_i X$.

<u>5.8 Remark.</u> In practice, i.e. for spectral sequences with <u>countable</u> groups $E_r^{s,t}$, Mittag-Leffler convergence is <u>equivalent</u> to complete convergence (see [Gray, p. 242]).

Chapter X. Cosimplicial spaces

§1. Introduction

In this chapter we

(i) lay the foundation for a homotopy theory of cosimplicial spaces, i.e. we show that it is possible to define, for cosimplicial spaces, notions of function space, weak equivalence, cofibration and fibration, which satisfy Quillen's axioms for a closed simplicial model category (see Ch. VIII, 4.9), and then

(ii) combine this with the results of Chapter IX and obtain, for every cosimplicial space, an extended homotopy spectral sequence, which is an important tool in our study of the R-completion of a space in Part I.

In slightly more detail:

§2 contains a definition of cosimplicial spaces and a few examples.

§3 Here we define a notion of function space and discuss the important special case of the total space of a cosimplicial space, which is a kind of codiagonal.

§4 deals with the notions of weak equivalence, cofibration and fibration, and the closely related notions of cofibrant and fibrant cosimplicial spaces. A (for Part I of these notes) important example of such fibrant objects are the so-called grouplike cosimplicial spaces.

§5 is devoted to the verification of Quillen's axioms.

§6 Here we construct, for every cosimplicial (pointed) space, an *extended homotopy spectral sequence*, which, under suitable circumstances, converges to the *homotopy groups of the total space*.

§7 contains a *cosimplicial description of the* E_2-*term* of the spectral sequence, which is convenient for the applications in Chapter I.

Notation. We will work mainly in the categories $\mathcal{S}$ *of spaces* and $\mathcal{S}_*$ *of pointed spaces*.

§2. Cosimplicial spaces

This section contains a definition of cosimplicial spaces and, more generally, of cosimplicial objects over an arbitrary category, and a few examples.

2.1 Cosimplicial objects and maps. For a category $\mathcal{D}$, the category $c\mathcal{D}$ of cosimplicial objects over $\mathcal{D}$ is defined as follows. An object $\underset{\sim}{X} \in c\mathcal{D}$ consists of

(i) for every integer $n \geq 0$, an object $\underset{\sim}{X}^n \in \mathcal{D}$,

(ii) for every pair of integers (i,n) with $0 \leq i \leq n$, coface and codegeneracy maps

$$d^i\colon \underset{\sim}{X}^{n-1} \longrightarrow \underset{\sim}{X}^n \qquad \text{and} \qquad s^i\colon \underset{\sim}{X}^{n+1} \longrightarrow \underset{\sim}{X}^n \qquad \in \mathcal{D}$$

satisfying the cosimplicial identities (which are dual to the simplicial identities (Ch. VIII, 2.1)):

$$\begin{aligned}
d^j d^i &= d^i d^{j-1} && \text{for } i < j \\
s^j d^i &= d^i s^{j-1} && \text{for } i < j \\
&= \mathrm{id} && \text{for } i = j,\ j+1 \\
&= d^{i-1} s^j && \text{for } i > j+1 \\
s^j s^i &= s^{i-1} s^j && \text{for } i > j
\end{aligned}$$

Similarly a cosimplicial map $f\colon \underset{\sim}{X} \to \underset{\sim}{Y} \in c\mathcal{D}$ consists of maps

$$f\colon \underset{\sim}{X}^n \longrightarrow \underset{\sim}{Y}^n \in \mathcal{D}$$

which commute with the coface and codegeneracy maps. A cosimplicial object (map) over $\mathcal{D}$ thus corresponds to a simplicial object (map) over the dual category $\mathcal{D}^*$ (Ch. VIII, 2.1).

2.2 Examples

(i) The cosimplicial standard simplex

$$\underset{\sim}{\Delta} \in c\mathcal{S}$$

i.e. the cosimplicial space which in codimension n consists of the standard n-simplex $\Delta[n] \in \mathcal{S}$ and for which the coface and codegeneracy maps are the standard maps (Ch. VIII, 2.9 and 2.11).

$$\Delta[n-1] \xrightarrow{d^j} \Delta[n] \qquad \text{and} \qquad \Delta[n+1] \xrightarrow{s^j} \Delta[n].$$

(ii) For $X, Y \in \mathcal{S}_*$ one can form the cosimplicial pointed space $\underset{\sim}{\text{hom}}_*(X,Y) \in c\mathcal{S}_*$, where

$$\underset{\sim}{\text{hom}}_*(X,Y)^n_k = \{\text{pointed maps } X_n \longrightarrow Y_k\}.$$

(iii) Our key example of a cosimplicial space is the cosimplicial resolution $\underset{\sim}{R}X$ of a space X with respect to a ring R (Ch. I, 4.1)

(iv) A diagram

$$X \xrightarrow{f} B \xleftarrow{g} Y \qquad \in \mathcal{S}$$

gives rise to a cosimplicial space $X \underset{\sim}{\times}_B Y$ with

$$(X \underset{\sim}{\times}_B Y)^n = X\times B\times \cdots \times B\times Y \qquad \text{(n copies of B)}$$

$$d^i(x,b_1,\ldots,b_n,y) = \begin{cases} (x,fx,b_1,\ldots,b_n,y) & i = 0 \\ (x,b_1,\ldots,b_i,b_i,\ldots,b_n,y) & 1 \le i \le n \\ (x,b_1,\ldots,b_n,gy,y) & i = n+1 \end{cases}$$

$$s^i(x,b_1,\ldots,b_n,y) = (x,b_1,\ldots,b_i,b_{i+2},\ldots,b_n,y) \quad 0 \le i \le n-1$$

This example was used by [Rector (EM)] in his geometric construction of the Eilenberg-Moore spectral sequence.

§3. The total space of a cosimplicial space

We will now associate with every cosimplicial space a very useful space, its *total space*. This is a kind of *codiagonal*; it is, in some sense, dual to the diagonal of a simplicial space (a simplicial space is a double-simplicial set).

Total spaces are a special case of

3.1 Function spaces. Just as we defined (in Ch. VIII, 4.7), for $X, Y \in \mathcal{S}$, the function space $\hom(X,Y) \in \mathcal{S}$, so we now define, for $\underset{\sim}{X}, \underset{\sim}{Y} \in c\mathcal{S}$, the *function space*

$$\hom(\underset{\sim}{X},\underset{\sim}{Y}) \in \mathcal{S}$$

as the space of which the n-simplices are the maps

$$\Delta[n] \times \underset{\sim}{X} \longrightarrow \underset{\sim}{Y} \quad \in \quad c\mathcal{S}$$

with as faces and degeneracies the compositions

$$\Delta[n-1] \times \underset{\sim}{X} \xrightarrow{d^i \times \underset{\sim}{X}} \Delta[n] \times \underset{\sim}{X} \longrightarrow \underset{\sim}{Y}$$

$$\Delta[n+1] \times \underset{\sim}{X} \xrightarrow{s^i \times \underset{\sim}{X}} \Delta[n] \times \underset{\sim}{X} \longrightarrow \underset{\sim}{Y}$$

As already said, a very useful example of a function space is

3.2 The total space of a cosimplicial space. For $\underset{\sim}{X} \in c\mathcal{S}$ we define its *total space* $\operatorname{Tot} \underset{\sim}{X}$ or $\operatorname{Tot}_\infty \underset{\sim}{X}$ by (2.2)

$$\operatorname{Tot}_\infty \underset{\sim}{X} = \operatorname{Tot} \underset{\sim}{X} = \hom(\underset{\sim}{\Delta}, \underset{\sim}{X}) \qquad \in \mathcal{S}$$

and note that the total space can be considered as an inverse limit

$$\operatorname{Tot} \underset{\sim}{X} = \lim_{\leftarrow} \operatorname{Tot}_s \underset{\sim}{X}$$

where

$$\operatorname{Tot}_s \underset{\sim}{X} = \hom(\underset{\sim}{\Delta}^{[s]}, \underset{\sim}{X}) \qquad \varepsilon \mathscr{S}$$

and $\underset{\sim}{\Delta}^{[s]} \subset \underset{\sim}{\Delta}$ denotes the simplicial s-skeleton, i.e. $\underset{\sim}{\Delta}^{[s]}$ consists in codimension n of the s-skeleton (Ch. VIII, 2.13) of $\Delta[n]$.

If $\underset{\sim}{X} \varepsilon c\mathscr{S}$ is augmented, i.e. comes with an augmentation map

$$d^0 \colon \underset{\sim}{X}^{-1} \longrightarrow \underset{\sim}{X}^0 \qquad \varepsilon \mathscr{S}$$

such that $d^0 d^0 = d^1 d^0 \colon \underset{\sim}{X}^{-1} \to \underset{\sim}{X}^1$, then this augmentation map obviously induces maps

$$\phi \colon \underset{\sim}{X}^{-1} \longrightarrow \operatorname{Tot}_s \underset{\sim}{X} \quad \varepsilon \mathscr{S} \qquad -1 \leq s \leq \infty$$

which are compatible with the maps between the $\operatorname{Tot}_s \underset{\sim}{X}$.

3.3 Examples

(i) For X, Y $\varepsilon \mathscr{S}_*$, the functors Tot and Tot_s give rise to the usual pointed function spaces (2.2 (ii) and Ch. VIII, 4.8)

$$\operatorname{Tot} \underset{\sim}{\hom_*}(X,Y) = \hom_*(X,Y) \qquad \operatorname{Tot}_s \underset{\sim}{\hom_*}(X,Y) = \hom_*(X^{[s]},Y)$$

(ii) The R-completion $R_\infty X$ of a space X with respect to a ring R, which is (Ch. I, §4) defined by (see 2.2 (ii))

$$R_\infty X = \operatorname{Tot} \underset{\sim}{R}X$$

(iii) Given a diagram

$$X \xrightarrow{f} B \xleftarrow{g} Y \qquad \in \mathcal{S}$$

one can form the diagram

$$X \xrightarrow{f} B \xleftarrow{\hom(d^1,B)} \hom(\Delta[1],B) \xrightarrow{\hom(d^0,B)} B \xleftarrow{g} Y$$

and verify easily that (see 2.2 (iv))

$$\operatorname{Tot}(X \times_{\underset{\sim}{B}} Y) \simeq X \times_B \hom(\Delta[1],B) \times_B Y$$

Thus, <u>if B is fibrant and f and g are fibrations, then the natural map</u>

$$X \times_B Y \longrightarrow \operatorname{Tot}(X \times_{\underset{\sim}{B}} Y)$$

<u>is a homotopy equivalence</u>.

We end with another example of a function space.

3.4 The maximal augmentation. Let $\underset{\sim}{*} \in c\mathcal{S}$ denote the cosimplicial space with one element in each bi-dimension. Then it is not hard to see that, for $X \in c\mathcal{S}$, the space $\hom(\underset{\sim}{*}, \underset{\sim}{X})$ is naturally isomorphic to the <u>maximal augmentation</u> of $\underset{\sim}{X}$, i.e. the subspace of $\underset{\sim}{X}^0$ which consists of the simplices $x \in \underset{\sim}{X}^0$ for which $d^0x = d^1x$.

§4. Weak equivalences, cofibrations and fibrations

In this section we define for cosimplicial spaces notions of weak equivalences, cofibrations and fibrations, which (as will be shown in §5) have all the "usual" properties.

4.1 Weak equivalences. A map $f: \underset{\sim}{X} \to \underset{\sim}{Y} \in c\mathscr{S}$ will be called a weak equivalence if the maps

$$f: \underset{\sim}{X}^n \longrightarrow \underset{\sim}{Y}^n \in \mathscr{S} \qquad n \geq 0$$

are all weak equivalences.

4.2 Cofibrations. A map $i: \underset{\sim}{A} \to \underset{\sim}{B} \in c\mathscr{S}$ will be called a cofibration if it is 1-1 and induces an isomorphism on the maximal augmentation (3.4). This readily implies that every simplex $b \in \underset{\sim}{B}$ which is not in the image of i can uniquely be written in the form

$$b = d^{j_m} \dots d^{j_1} b'$$

where $j_m > \dots > j_i$ and b' is not a coface.

We call an object $\underset{\sim}{B} \in c\mathscr{S}$ unaugmentable or cofibrant if the map $\underset{\sim}{\emptyset} \to \underset{\sim}{B}$ is a cofibration ($\underset{\sim}{\emptyset}$ denotes the empty cosimplicial space), i.e. if the maximal augmentation of $\underset{\sim}{B}$ is empty.

4.3 Examples

(i) The cosimplicial standard simplex $\underset{\sim}{\Delta}$ (2.2) and its simplicial skeletons $\underset{\sim}{\Delta}^{[s]}$ (3.2) are unaugmentable.

(ii) The inclusion maps $\underset{\sim}{\Delta}^{[s]} \to \underset{\sim}{\Delta}^{[n]}$ $(s \leq n)$ are cofibrations and so are the maps $\underset{\sim}{*} \to \underset{\sim}{\Delta}^{[n]}/\underset{\sim}{\Delta}^{[s]}$.

4.4 Remark. We could now define a map $\underset{\sim}{X} \to \underset{\sim}{Y} \in c\mathcal{S}$ to be a fibration if it has the right lifting property with respect to all cofibrations which are weak equivalences, i.e. if for every (commutative) solid arrow diagram in $c\mathcal{S}$

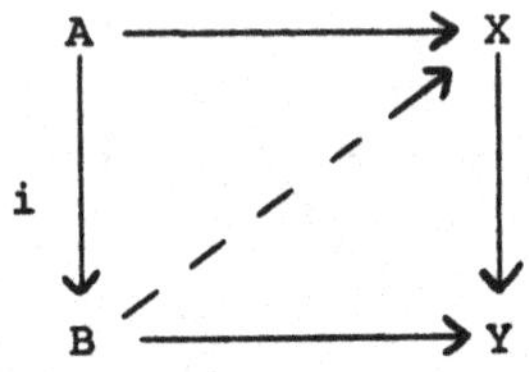

where i is a cofibration which is a weak equivalence, the dotted arrow exists. Instead we shall give an equivalent, but more explicit definition using

4.5 Matching spaces. For $\underset{\sim}{X} \in c\mathcal{S}$ and $n \geq -1$, we construct a matching space $M^n\underset{\sim}{X} \in \mathcal{S}$, which is, roughly speaking, the cosimplicial analogue of "the set of the (n+1)-simplices of the n-skeleton of a simplicial set". It consists of the simplices

$$(x^0,\ldots,x^n) \quad \in \quad \underset{\sim}{X}^n \times \cdots \times \underset{\sim}{X}^n$$

for which $s^i x^j = s^{j-1}x^i$ whenever $0 \leq i < j \leq n$, and it comes with a natural map

$$s\colon \underset{\sim}{X}^{n+1} \longrightarrow M^n\underset{\sim}{X} \quad \in \quad \mathcal{S}$$

given by $x \to (s^0x,\ldots,s^nx)$ for all $x \in \underset{\sim}{X}^{n+1}$

Clearly

$$M^{-1}\underset{\sim}{X} = * \qquad \text{and} \qquad M^0\underset{\sim}{X} = \underset{\sim}{X}^0$$

We now define

4.6 Fibrations. A map $f: \underset{\sim}{X} \to \underset{\sim}{Y} \in c\mathcal{S}$ will be called a *fibration* if the maps

$$(f,s): \underset{\sim}{X}^{n+1} \longrightarrow \underset{\sim}{Y}^{n+1} \times_{M^n\underset{\sim}{Y}} M^n\underset{\sim}{X} \quad \in \mathcal{S} \qquad n \geq -1$$

are all fibrations.

Similarly we say that $\underset{\sim}{X} \in c\mathcal{S}$ is *fibrant*, if $\underset{\sim}{X} \to \underset{\sim}{*}$ is a fibration, i.e. if the maps

$$s: \underset{\sim}{X}^{n+1} \longrightarrow M^n\underset{\sim}{X} \quad \in \mathcal{S} \qquad n \geq -1$$

are all fibrations.

4.7 Examples

(i) If $X \to X' \in \mathcal{S}_*$ is a cofibration and $Y \in \mathcal{S}_*$ is fibrant, then the induced map (2.2 (ii))

$$\underset{\sim}{\text{hom}}_*(X',Y) \longrightarrow \underset{\sim}{\text{hom}}_*(X,Y) \quad \in \quad c\mathcal{S}$$

is a fibration.

(ii) If $Y \to Y' \in \mathcal{S}_*$ is a fibration and $X \in \mathcal{S}_*$, then the induced map (2.2 (ii))

$$\underset{\sim}{\text{hom}}_*(X,Y) \longrightarrow \underset{\sim}{\text{hom}}_*(X,Y') \quad \in \quad c\mathcal{S}$$

is also a fibration.

Another (for our purposes important) example is

4.8 Grouplike cosimplicial spaces. We call an object $\underset{\sim}{X} \in c\mathcal{S}$

grouplike if, for all $n \geq 0$, the space X^n is a simplicial group (i.e. a simplicial object over the category of groups) and the operators d^i (except d^0) and all operators s^i are homomorphisms of simplicial groups.

Grouplike objects have the following useful properties:

4.9 Proposition

(i) Every "homomorphism" $f: X \to Y \in c\mathcal{S}$ of grouplike objects, which is onto, is a fibration, and hence

(ii) Every grouplike object is fibrant

Proof. This follows from the fact that the maps

$$(f,s): X^{n+1} \longrightarrow Y^{n+1} \times_{M^nY} M^nX$$

are epimorphisms of simplicial groups, and hence [May, p. 70] fibrations. This, in turn, is a consequence of the fact that the maps $s: V^{n+1} \to M^nV$ have a natural (simplicial) cross section when V is grouplike. The proof of this last statement is very similar to the proof that every simplicial group is fibrant [May, p. 67], but uses codegeneracies instead of **faces and cofaces** instead of degeneracies.

4.10 Examples

(i) Every cosimplicial simplicial group is grouplike and hence fibrant

(ii) The cosimplicial resolution RX of a space X with respect to a ring R (2.2 (iii) and Ch. I, §4) is fibrant, because (Ch. I, 2.2) every choice of a base point $* \in X$ makes RX grouplike.

§5. Cosimplicial spaces form a closed simplicial model category

The purpose of this section is to prove that

5.1 The category $c\mathcal{S}$ is a closed simplicial model category, i.e. the notions of function space, weak equivalence, cofibration and fibration in the category $c\mathcal{S}$, which were defined in §3 and §4, satisfy the axioms CM1-5 and SM7 of [Quillen (HA), p. II, 2.2 and (RH), p. 233].

The axioms CM1-5 for a closed model category were listed in Ch. VIII, 3.5 and involve only the notions of weak equivalence, cofibration and fibration, while axiom SM7 relates the notion of function space with the others as follows.

SM7. If $i: \underset{\sim}{A} \to \underset{\sim}{B}$ is a cofibration and $p: \underset{\sim}{X} \to \underset{\sim}{Y}$ is a fibration, then the map

$$(i,p): \hom(\underset{\sim}{B},\underset{\sim}{X}) \longrightarrow \hom(\underset{\sim}{A},\underset{\sim}{X}) \times_{\hom(\underset{\sim}{A},\underset{\sim}{Y})} \hom(\underset{\sim}{B},\underset{\sim}{Y}) \quad \in \mathcal{S}$$

is a fibration, which is a weak equivalence if either i or p is a weak equivalence.

Before proving this we mention a useful consequence.

5.2 Proposition

(i) If $f: \underset{\sim}{X} \to \underset{\sim}{Y} \in c\mathcal{S}$ is a weak equivalence, with $\underset{\sim}{X}$ and $\underset{\sim}{Y}$ fibrant, and $\underset{\sim}{A} \in c\mathcal{S}$ is cofibrant, then f induces a homotopy equivalence

$$\hom(\underset{\sim}{A},\underset{\sim}{X}) \simeq \hom(\underset{\sim}{A},\underset{\sim}{Y}) \quad \in \mathcal{S}$$

(ii) *If* $g: \underset{\sim}{A} \to \underset{\sim}{B} \in c\mathcal{S}$ *is a weak equivalence, with* $\underset{\sim}{A}$ *and* $\underset{\sim}{B}$ *cofibrant, and* $\underset{\sim}{X} \in c\mathcal{S}$ *is fibrant, then* g *induces a homotopy equivalence*

$$\hom(\underset{\sim}{B},\underset{\sim}{X}) \simeq \hom(\underset{\sim}{A},\underset{\sim}{X}) \quad \in \mathcal{S}$$

Proof. It follows from 5.1, that f can be factored f = pi, where p is a fibration, i is a cofibration, and both are weak equivalences; moreover hom $(\underset{\sim}{A},p)$ is a weak equivalence. Now, by [Quillen (HA), p. II, 2.5] i is a strong deformation retract map. As hom $(\underset{\sim}{A},-)$ preserves the simplicial homotopy relation, this implies that hom $(\underset{\sim}{A},i)$ is a weak equivalence. This proves part (i).

The proof of part (ii) is similar.

Proof of 5.1. We will only prove the "difficult" lifting and factorization axioms CM4 and CM5. The axioms CM1, CM2 and CM3 are easy and will be left to the reader, while SM7 follows from [Quillen (HA), p. II, 2.3 axiom SM7b] which is obvious in our case. First a

5.3 Lemma. *A fibration* $f: \underset{\sim}{X} \to \underset{\sim}{Y} \in c\mathcal{S}$ *is a weak equivalence if and only if the maps*

$$(f,s): \underset{\sim}{X}^{n+1} \longrightarrow \underset{\sim}{Y}^{n+1} \times_{M^n\underset{\sim}{Y}} M^n\underset{\sim}{X} \quad \in \mathcal{S} \qquad n \geq -1$$

are all weak equivalences.

Proof. For $n \geq -1$ and $-1 \leq k \leq n$ let $M_k^n\underset{\sim}{X}$ consist of the simplices

$$(x^0,\ldots,x^k) \quad \in \quad \underset{\sim}{X}^n \times \cdots \times \underset{\sim}{X}^n$$

for which $s^i x^j = s^{j-1} x^i$ whenever $0 \le i < j \le k$. Clearly $M^n_n \underset{\sim}{X} = M^n \underset{\sim}{X}$ and if $k = -1$ or $n = -1$, then $M^n_k \underset{\sim}{X} = *$.

The lemma now follows by an inductive argument since the obvious map

$$\underset{\sim}{Y}^{n+1} \times_{M^n_{k+1}\underset{\sim}{Y}} M^n_{k+1}\underset{\sim}{X} \longrightarrow \underset{\sim}{Y}^{n+1} \times_{M^n_k \underset{\sim}{Y}} M^n_k \underset{\sim}{X}$$

is a fibration induced from the obvious map

$$\underset{\sim}{X}^n \longrightarrow \underset{\sim}{Y}^n \times_{M^{n-1}_k \underset{\sim}{Y}} M^{n-1}_k \underset{\sim}{X}.$$

Proof of CM5. For $m, n \ge 0$, let

$$\underset{\sim}{\Delta}[{m \atop n}] \quad \varepsilon \quad c\mathscr{A}$$

be the object freely generated by a simplex $i^m_n \in \underset{\sim}{\Delta}[{m \atop n}]^m_n$, and let $\dot{\underset{\sim}{\Delta}}[{m \atop n}] \subset \underset{\sim}{\Delta}[{m \atop n}]$ be the sub-object generated by the simplices

$$s^j i^m_n \qquad 0 \le j < m$$

$$d_k i^m_n \qquad 0 \le k \le n, \quad n > 0 .$$

Then the inclusion $\dot{\underset{\sim}{\Delta}}[{m \atop n}] \to \underset{\sim}{\Delta}[{m \atop n}]$ is a cofibration and, by 5.3, a map $p: \underset{\sim}{X} \to \underset{\sim}{Y} \in c\mathscr{A}$ has the right lifting property (4.4) with respect to all the maps $\dot{\underset{\sim}{\Delta}}[{m \atop n}] \to \underset{\sim}{\Delta}[{m \atop n}]$ if and only if p is a fibration and a weak equivalence. Thus any map $f \in c\mathscr{A}$ may be factored $f = pi$, where p is a fibration and a weak equivalence, and where i is a (possibly transfinite) composition of co-base extension of maps $\dot{\underset{\sim}{\Delta}}[{m \atop n}] \to \underset{\sim}{\Delta}[{m \atop n}]$.

Similarly, for $0 \le k \le n$ and $n > 0$, let $\underset{\sim}{\Delta}[{m \atop n,k}] \subset \underset{\sim}{\Delta}[{m \atop n}]$ be the sub-object generated by the simplices

$$s^j i^m_n \qquad\qquad 0 \le i < m$$

$$d_j i^m_n \qquad\qquad 0 \le j \le n, \quad j \ne k$$

Then the inclusion $\underset{\sim}{\Delta}\left[{m \atop n,k}\right] \to \underset{\sim}{\Delta}\left[{m \atop n}\right]$ is a cofibration and a weak equivalence. Thus any map $f \in c\mathcal{A}$ may be factored $f = pi$, where p is a fibration and where i is a (possibly transfinite) composition of co-base extensions of maps $\underset{\sim}{\Delta}\left[{m \atop n,k}\right] \to \underset{\sim}{\Delta}\left[{m \atop n}\right]$.

<u>Proof of CM4</u>. The case where p is a weak equivalence is easy, since any cofibration is a (possibly transfinite) composition of co-base extensions of maps $\dot{\underset{\sim}{\Delta}}\left[{m \atop n}\right] \to \underset{\sim}{\Delta}\left[{m \atop n}\right]$.

Now suppose i is a weak equivalence. Then, by the proof of CM5, there is a factorization $i = p'i'$ such that p' is a fibration and a weak equivalence and i' is a (possibly transfinite) composition of co-base extensions of maps $\underset{\sim}{\Delta}\left[{m \atop n,k}\right] \to \underset{\sim}{\Delta}\left[{m \atop n}\right]$. Since i' has the left lifting property (Ch. VIII, 3.4) with respect to fibrations, the desired result now follows easily.

§6. The homotopy spectral sequence of a cosimplicial space

Combining the preceding results with those of Chapter IX, §4, we construct, for every cosimplicial pointed space, a pointed tower of fibrations and hence an extended homotopy spectral sequence. Under suitable circumstances this spectral sequence converges to the homotopy groups of the total space of the cosimplicial space.

6.1 The homotopy spectral sequence of a cosimplicial pointed space. Let $\underset{\sim}{X} \in c\mathcal{S}_*$. If $\underset{\sim}{X}$ is fibrant, then (3.2, 4.3 and 5.1) $\{\mathrm{Tot}_n \underset{\sim}{X}\}$ is a pointed tower of fibrations, and we define the (extended) homotopy spectral sequence $\{E_r^{s,t} \underset{\sim}{X}\}$ by (Ch. IX, 4.2)

$$E_r^{s,t} \underset{\sim}{X} = E_r^{s,t}\{\mathrm{Tot}_n \underset{\sim}{X}\}.$$

Otherwise we choose (5.1) a weak equivalence $\underset{\sim}{X} \to \underset{\sim}{Y} \in c\mathcal{S}_*$ such that $\underset{\sim}{Y}$ is fibrant and put

$$E_r^{s,t} \underset{\sim}{X} = E_r^{s,t} \{\mathrm{Tot}_n \underset{\sim}{Y}\}.$$

Under favorable conditions (Ch. IX, 5.4) this spectral sequence $\{E_r \underset{\sim}{X}\}$ converges to $\pi_*\mathrm{Tot}\, \underset{\sim}{X}$.

That $E_r^{s,t} \underset{\sim}{X}$ is indeed well-defined (and, of course, natural) follows readily from 5.1 and the following

6.2 Description of the E_1-term. *Let $\underset{\sim}{X} \in \mathcal{S}_*$. Then there exist natural isomorphisms*

$$E_1^{s,t}\underset{\sim}{X} \approx \pi_t \underset{\sim}{X}^s \cap \ker s^0 \cap \ldots \cap \ker s^{s-1} \qquad t \geq s \geq 0 .$$

This, in turn, follows readily from

<u>6.3 Proposition.</u> <u>Let</u> $X \in c\mathcal{S}_*$ <u>be fibrant.</u> <u>Then, for all n,</u>

<u>(i) the fibre of the map</u> $\mathrm{Tot}_n X \to \mathrm{Tot}_{n-1} X$ <u>is the pointed function space</u> $\hom_*(S^n, NX^n)$ (Ch. IX, 3.2 and Ch. VIII, 2.12) <u>where</u>

$$NX^n = \ker (X^n \xrightarrow{s} M^{n-1}X)$$

$$= X^n \cap \ker s^0 \cap \dots \cap \ker s^{s-1} .$$

<u>(ii) for each</u> $i \geq 0$

$$\pi_i NX^n = \ker (\pi_i X^n \xrightarrow{s} M^{n-1}\pi_i X)$$

$$= \pi_i X^n \cap \ker s^0 \cap \dots \cap \ker s^{s-1} .$$

<u>Proof.</u> Part (i) is obvious, and for part (ii) it suffices to show that the obvious map $\pi_i M_k^n X \to M_k^n \pi_i X$ (see proof of 3.2) is an isomorphism for all i, k and n. This follows inductively from the fact that the maps $\pi_i X^n \to M_k^{n-1}\pi_i X$ are onto for $i \geq 1$ (4.9) and that there are pull backs

$$\begin{array}{ccc} M_{k+1}^n X & \longrightarrow & X^n \\ \downarrow & & \downarrow \\ M_k^n X & \longrightarrow & M_k^{n-1} X \end{array} \qquad \begin{array}{ccc} M_{k+1}^n \pi_i X & \longrightarrow & \pi_i X^n \\ \downarrow & & \downarrow \\ M_k^n \pi_i X & \longrightarrow & M_k^{n-1} \pi_i X \end{array}$$

<u>6.4 Remark.</u> The above description of the E_1-term (6.2) implies, in view of the main results of [Bousfield-Kan (SQ), §7 and §10] that

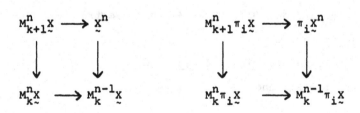

(i) the spectral sequence $\{E_r^{s,t} X\}$ defined above coincides in dimensions ≥ 1 with the spectral sequence of [Bousfield-Kan (SQ), §7] and hence

(ii) if $X \in \mathcal{S}_*$ and $\underset{\sim}{R}X \in c\mathcal{S}_*$ is the cosimplicial resolution of X with respect to a ring R (2.2 (iii) and Ch. I, 4.1), then $\{E_r^{s,t}\underset{\sim}{R}X\}$ coincides in dimension ≥ 1 with the <u>homotopy spectral sequence</u> $\{E_r^{s,t}(X;R)\}$ <u>of X with coefficients in R</u> of [Bousfield-Kan (HS)];

(iii) for $\underset{\sim}{X}, \underset{\sim}{Y} \in c\mathcal{S}_*$ there is a natural <u>pairing</u> (at least in dimensions ≥ 1)

$$E_r^{s,t}\underset{\sim}{X} \otimes E_r^{s,t}\underset{\sim}{Y} \longrightarrow E_r^{s+s',t+t'}(\underset{\sim}{X} \wedge \underset{\sim}{Y}).$$

(iv) if $\underset{\sim}{X}$ is a cosimplicial simplicial abelian group, then $\{E_r^{s,t}\underset{\sim}{X}\}$ is (part of) the usual <u>spectral sequence of the second quadrant double chain complex</u> obtained by "doubly normalizing" $\underset{\sim}{X}$.

§7. A cosimplicial description of the E_2-term

We end this chapter with a cosimplicial description for the E_2-term of the spectral sequence of §6. For this we need a notion of

7.1 Cohomotopy groups (and pointed sets). For a cosimplicial abelian group B we define its cohomotopy groups $\pi^s B$ by

$$\pi^s B = H^s(B,d) \qquad s \geq 0$$

where B is considered as a cochain complex with coboundary map $d = \Sigma(-1)^i d^i$, or equivalently

$$\pi^s B = H^s(NB,d)$$

where (NB,d) is the normalized cochain complex, i.e. the subcomplex given by

$$NB^n = B^n \cap \ker s^0 \cap \ldots \cap \ker s^{s-1}.$$

More generally, for a cosimplicial group B, one can still construct a cohomotopy group $\pi^0 B$ by

$$\pi^0 B = \{b \in B \mid d^0 b = d^1 b\}$$

and a pointed cohomotopy set $\pi^1 B$ as the orbit set of

$$ZB^1 = \{b \in NB^1 \mid (d^0 b)(d^1 b)^{-1}(d^2 b) = * \}$$

under the group action $B^0 \times ZB^1 \to ZB^1$ given by the formula $(b,b') \to (d^1 b)b'(d^0 b)^{-1}$.

And finally, for a cosimplicial pointed set B, the above construction still yields a pointed cohomotopy set $\pi^0 B$.

Then it is not hard to prove the following

7.2 Description of the E_2-term. Let $\underset{\sim}{X} \in c\mathcal{S}_*$. Then there exist natural isomorphisms

$$E_2^{s,t}\underset{\sim}{X} \approx \pi^s \pi_t \underset{\sim}{X} \qquad t \geq s \geq 0.$$

We now combine this description of E_2 with Ch. IX, 5.1 and 5.2 and get

7.3 Connectivity lemma. Let $k \geq 0$ and let $\underset{\sim}{X} \in c\mathcal{S}_*$ be fibrant and be such that $\pi^s \pi_t \underset{\sim}{X} = *$ for $0 \leq t-s \leq k$. Then

$$\varprojlim \pi_i \operatorname{Tot}_n \underset{\sim}{X} = * = \varprojlim{}^1 \pi_{i+1} \operatorname{Tot}_n \underset{\sim}{X} \qquad 0 \leq i \leq k$$

and hence $\operatorname{Tot} \underset{\sim}{X}$ is k-connected.

7.4 Mapping lemma. Let $\underset{\sim}{X}, \underset{\sim}{Y} \in c\mathcal{S}_*$ be fibrant, let $\pi^s \pi_t X = *$ for $t-s = 0$ and let $f: \underset{\sim}{X} \to \underset{\sim}{Y} \in c\mathcal{S}_*$ induce isomorphisms $\pi^s \pi_t \underset{\sim}{X} \approx \pi^s \pi_t \underset{\sim}{Y}$ for all $t-s \geq 0$. Then f induces a homotopy equivalence

$$\operatorname{Tot} \underset{\sim}{X} \simeq \operatorname{Tot} \underset{\sim}{Y} \quad \in \mathcal{S}$$

We end with some

7.5 Examples

(i) If $Y \in \mathcal{S}_*$ is fibrant, then so is $\underset{\sim}{\hom}_*(X,Y)$ (2.2) for all $X \in \mathcal{S}_*$. Moreover

$$\pi^s \pi_t \underset{\sim}{hom}_*(X,Y) \simeq \tilde{H}^s(X;\ \pi_t Y)$$

(where $\tilde{H}^s$ denotes reduced cohomology) and hence (3.3) $\{E_r \underset{\sim}{hom}_*(X,Y)\}$ provides a new construction for the well-known <u>spectral sequence of a pointed function space</u>.

(ii) If $X, B, Y \in \mathcal{S}_*$ are fibrant, then so is $X \underset{\sim}{\times}_B Y$ (2.2) for any diagram

$$X \xrightarrow{f} B \xleftarrow{g} Y \qquad \in \mathcal{S}_*$$

Moreover

$$\pi^s \pi_t (X \underset{\sim}{\times}_B Y) = * \qquad \text{for } s > 1$$

$$\pi^0 \pi_t (X \underset{\sim}{\times}_B Y) = \pi_t X \times_{\pi_t B} \pi_t Y$$

$$\pi^1 \pi_t (X \underset{\sim}{\times}_B Y) = \pi_t B \ /\ \text{action}$$

where the action of $\pi_t X \times \pi_t Y$ on $\pi_t B$ is given by the formula $(u,v)b = (g_* v) b (f_* u)^{-1}$. Hence (3.3) $\{E_r (X \underset{\sim}{\times}_B Y)\}$ provides a new construction for the familiar <u>Mayer-Vietoris sequence of a fibre square</u>.

Chapter XI. Homotopy inverse limits

§1. Introduction

It is well known that, in general, inverse limits do not exist in the homotopy category of spaces. In this chapter we will, however:

(i) discuss a notion of homotopy inverse limits which gets around this difficulty, and

(ii) show that, up to homotopy, the R-completion of Part I of these notes can be considered as such a homotopy inverse limit.

In more detail: if I is a small category, $\mathcal{S}$ the category of spaces, and $\mathcal{S}^I$ the category of I-diagrams in $\mathcal{S}$, then the homotopy inverse limit is a certain functor

$$\underleftarrow{\text{holim}}: \mathcal{S}^I \longrightarrow \mathcal{S}$$

which satisfies the homotopy lemma:

If $f: \underline{X} \to \underline{X}' \in \mathcal{S}^I$ is a map, of which each "component" is a homotopy equivalence between fibrant spaces, then f induces a homotopy equivalence

$$\underleftarrow{\text{holim}}\, f: \underleftarrow{\text{holim}}\, \underline{X} \simeq \underleftarrow{\text{holim}}\, \underline{X}' \qquad \in \mathcal{S}$$

Moreover $\underleftarrow{\text{holim}}$ has the "desired" homotopy type in familiar examples, such as:

(i) If $\underline{X}$ is a tower of fibrations, then $\underleftarrow{\text{holim}}\, \underline{X}$ is homotopy equivalent to $\underleftarrow{\lim}\, \underline{X}$.

(ii) If $\underline{X}$ is a fibrant cosimplicial space, then $\operatorname{holim}_{\leftarrow} \underline{X}$ is homotopy equivalent to the total space $\operatorname{Tot} \underline{X}$.

(iii) If $\underline{A}$ is an I-diagram of abelian groups and $K(\underline{A},n) \in \mathcal{S}^I$ is the corresponding I-diagram of Eilenberg-MacLane spaces, then

$$\pi_i \operatorname{holim}_{\leftarrow} K(\underline{A},n) \approx \lim_{\leftarrow}{}^{n-i} \underline{A} \qquad 0 \leq i \leq n$$

$$= 0 \qquad \text{otherwise}$$

where $\lim\limits_{\leftarrow}{}^{n-i}$ is the (n-i)-th derived functor of $\lim\limits_{\leftarrow}$.

We also construct, for homotopy inverse limits a (homotopy) spectral sequence, which generalizes the one for cosimplicial spaces of Chapter X, §6; and we prove a cofinality theorem.

The chapter is organized as follows:

§2 and §3 After some preparations in §2, we give in §3 several, of course equivalent, descriptions of the homotopy inverse limit functor $\operatorname{holim}_{\leftarrow}$.

§4 contains some examples, and a proof of the fact that, for a fibrant cosimplicial space, the homotopy inverse limit and the total space have the same homotopy type.

We also observe that the definition of homotopy inverse limits in $\mathcal{S}$ can be generalized to any category $\mathcal{C}$ which

(i) has (ordinary) inverse limits, and

(ii) has natural "function objects" $\hom(K,X) \in \mathcal{C}$, for $K \in \mathcal{S}$ and $X \in \mathcal{C}$.

Thus homotopy inverse limits can, for instance, be defined in the categories $\mathcal{S}_*$ of pointed spaces, $\mathcal{T}$ of topological spaces, and

$\mathcal{T}_*$ of pointed topological spaces; however, nothing really new happens in $\mathcal{S}_*$, $\mathcal{T}$ or $\mathcal{T}_*$.

§5 Here we show that the functor $\underleftarrow{\text{holim}}: \mathcal{S}^I \to \mathcal{S}$ admits a factorization through the category $c\mathcal{S}$ of cosimplicial spaces

$$\mathcal{S}^I \xrightarrow{\prod^*} c\mathcal{S} \xrightarrow{\text{Tot}} \mathcal{S} \quad .$$

This implies that one can use cosimplicial methods to study homotopy inverse limit spaces.

§6 and §7 contain an example of this. We construct for a pointed diagram of fibrant spaces $\underline{X}$, a spectral sequence $\{E_r\underline{X}\}$ such that

(i) $\{E_r\underline{X}\}$ is usually closely related to $\pi_*\underleftarrow{\text{holim}}\ \underline{X}$,

(ii) $E_2^{s,t}\underline{X} \approx \underleftarrow{\lim}^s \pi_t\underline{X}$ for $0 \leq s \leq t$, where $\underleftarrow{\lim}^s$ denotes the s-th derived functor of the inverse limit functor for (abelian) groups, and

(iii) if $\underline{X}$ is a cosimplicial space, then, as one might expect, this spectral sequence coincides, from E_2 on, with the spectral sequence of Ch.X, §6.

§8 contains a brief discussion of an interpretation of homotopy inverse limits in terms of homotopy categories.

§9 and §10 Here we observe, that for certain large diagrams of spaces, one can, at least up to homotopy, talk of their homotopy inverse limits, and show that, as a consequence, the R-completion of Part I of these notes can, up to homotopy, be considered as a homotopy inverse limit of the "Artin-Mazur-like" diagram of "target spaces of X which admit a simplicial R-module structure". Our main tool is a cofinality theorem, which enables us to compare homotopy inverse limits for small diagrams of different "shapes".

Notation and terminology. In this chapter we will mainly work in the category $\mathcal{S}$ of spaces, except for §7, where we will also use the category $\mathcal{S}_*$ of pointed spaces.

We shall assume that the reader is familiar with ordinary inverse limits for diagrams in various categories (see [Kan (AF)] and [Artin-Mazur, Appendix]).

§2. Some spaces associated with a small category

In preparation for the definition of homotopy inverse limits (in §3) we discuss several useful spaces which one can associate with a small (i.e. the objects form a set) category. We start with

2.1 The underlying space of a small category. Let I be a small category. Then we will denote by the same symbol I its underlying space, i.e. the space of which an n-simplex is any sequence

$$u = (i_0 \xleftarrow{\alpha_1} \cdots \xleftarrow{\alpha_n} i_n) \qquad \in I$$

with faces and degeneracies given by

$$d_0 u = (i_1 \xleftarrow{\alpha_2} \cdots \xleftarrow{\alpha_n} i_n)$$

$$d_j u = (i_0 \xleftarrow{\alpha_1} \cdots \xleftarrow{\alpha_j\alpha_{j+1}} \cdots \xleftarrow{\alpha_n} i_n) \qquad 0 < j < n$$

$$d_n u = (i_0 \xleftarrow{\alpha_1} \cdots \xleftarrow{\alpha_{n-1}} i_{n-1})$$

$$s_j u = (i_0 \xleftarrow{\alpha_1} \cdots \xleftarrow{\alpha_j} i_j \xleftarrow{id} i_j \xleftarrow{\alpha_{j+1}} \cdots \xleftarrow{\alpha_n} i_n) \qquad 0 \le j \le n$$

Clearly, for small categories I and J, a functor $I \to J$ induces a simplicial map $I \to J$, and [Segal] it is not hard to see that:

(i) There is an obvious 1-1 correspondence between the functors $I \to J$ and the simplicial maps $I \to J$.

(ii) Moreover, a natural transformation of such functors corresponds to a homotopy

$$\Delta[1] \times I \longrightarrow J \qquad \in \mathcal{S} .$$

We also need

2.2 The (over) categories I/i and their underlying spaces. Given a small category and an object $i \in I$, one can form the (over) category I/i, which has as objects the maps

$$i \longleftarrow i_0 \qquad \in I$$

and as maps the appropriate commutative triangles. An n-simplex of the space I/i thus can be considered as a sequence

$$(i \xleftarrow{\alpha} i_0 \xleftarrow{\alpha_1} \cdots \xleftarrow{\alpha_n} i_n) \qquad \in I .$$

A map $\beta: i \to i' \in I$ induces a functor and hence a simplicial map

$$I/\beta: I/i \longrightarrow I/i' \qquad \in \mathcal{S}$$

and combining these for all β, one gets an I-diagram of spaces, i.e. a covariant functor

$$I/-: I \longrightarrow \mathcal{S}$$

which has the property:

2.3 Proposition. The correspondence

$$(i \xleftarrow{\alpha} i_0 \longleftarrow \cdots \longleftarrow i_n) \longrightarrow (i_0 \longleftarrow \cdots \longleftarrow i_n)$$

induces an isomorphism

$$\varinjlim I/- \approx I \qquad \in \mathcal{S} .$$

Another useful property is

2.4 Proposition. For every $i \in I$, the identity map of $I/i \in \mathcal{S}$ is homotopic to the composition

$$I/i \longrightarrow * \longrightarrow I/i \qquad \in \mathcal{S}$$

where the last map sends $*$ into $(i \xleftarrow{id} i)$.

This is immediate from 2.1.

2.5 Corollary. For every $i \in I$ and fibrant space $X \in \mathcal{S}$, the map $I/i \to *$ induces a weak equivalence

$$X \approx \hom(*,X) \longrightarrow \hom(I/i,X) \qquad \in \mathcal{S} .$$

2.6 Example. Let Δ be the category of the finite ordered sets

$$[n] = (0,\cdots,n) .$$

Then $\Delta/-$ can be considered as a cosimplicial space and it is not hard to see that

(i) $\Delta/-$ is cofibrant (i.e. unaugmentable),

(ii) the map

$$\Delta/- \longrightarrow \underset{\sim}{\Delta}$$

which sends every vertex $[i] \overset{\alpha}{\leftarrow} [i_0] \in \Delta/i$ into the vertex $\alpha(0) \in \Delta[i]$, is a cosimplicial map.

Moreover, in view of 2.4,

(iii) this map $\Delta/- \to \underset{\sim}{\Delta} \in c\mathcal{S}$ is a weak equivalence.

We end with the remark that, of course, 2.2 can be dualized. Thus one has

2.7 The (under) categories $I\backslash i$ and their underlying spaces. The definition is obvious. Note that these spaces give rise to a contravariant functor

$$I\backslash-: I \longrightarrow \mathcal{S} \quad .$$

§3. Homotopy inverse limits

It is convenient to define first

3.1 Function spaces for diagrams of spaces. Let I be a small category, let $\mathcal{S}^I$ be the category of I-diagrams over $\mathcal{S}$ (i.e. covariant functors $I \to \mathcal{S}$) and let $\underline{W},\underline{X} \in \mathcal{S}^I$. Then the function space

$$\hom(\underline{W},\underline{X}) \in \mathcal{S}$$

is the obvious (cf. Ch.VIII, §4) space of which the n-simplices are the maps

$$\Delta[n] \times \underline{W} \longrightarrow \underline{X} \in \mathcal{S}^I \quad .$$

Or equivalently, $\hom(W,X) \in \mathcal{S}$ is the difference kernel (i.e. inverse limit) of the maps

$$\prod_{i \in I} \hom(\underline{W}i,\underline{X}i) \underset{b}{\overset{a}{\rightrightarrows}} \prod_{i \xrightarrow{\gamma} i' \in I} \hom(\underline{W}i,\underline{X}i')$$

where a and b are respectively induced by

$$\hom(\underline{W}i,\underline{X}i) \xrightarrow{\underline{X}\gamma} \hom(\underline{W}i,\underline{X}i')$$

$$\hom(\underline{W}i',\underline{X}i') \xrightarrow{\underline{W}\gamma} \hom(\underline{W}i,\underline{X}i')$$

Now we can define

3.2 Homotopy inverse limits. Let I be a small category and let $\underline{X} \in \mathcal{S}^I$ be an I-diagram. The homotopy inverse limit of $\underline{X}$ then is the space $\operatorname{holim}_{\leftarrow} \underline{X}$ (or $\operatorname{holim}_{\leftarrow i} \underline{X}i$) defined by

$$\underset{\leftarrow}{\text{holim}}\ \underline{X} \quad = \quad \text{hom}(I/-,\underline{X}) \ \varepsilon\ \mathscr{S} \quad .$$

It is not hard to see that $\underset{\leftarrow}{\text{holim}}\ \underline{X}$ <u>is natural in</u> $\underline{X}$ <u>and</u> I: in particular, a map $f\colon \underline{X} \to \underline{X}' \ \varepsilon\ \mathscr{S}^I$ induces a map

$$\underset{\leftarrow}{\text{holim}}\ f\colon \underset{\leftarrow}{\text{holim}}\ \underline{X} \longrightarrow \underset{\leftarrow}{\text{holim}}\ \underline{X}' \qquad \varepsilon\ \mathscr{S}$$

and a functor $g\colon J \to I$ between small categories induces a functor $g^*\colon \mathscr{S}^I \to \mathscr{S}^J$, and hence a natural map

$$\underset{\leftarrow}{\text{holim}}\ g\colon \underset{\leftarrow}{\text{holim}}\ \underline{X} \longrightarrow \underset{\leftarrow}{\text{holim}}\ g^*\underline{X}$$

One can, of course, also obtain the functor $\underset{\leftarrow}{\text{holim}}$ using

3.3 An adjoint functor approach. <u>The functor</u>

$$\underset{\leftarrow}{\text{holim}}\colon \mathscr{S}^I \longrightarrow \mathscr{S}$$

<u>is right adjoint to the functor</u>

$$- \times (I/-)\colon \mathscr{S} \longrightarrow \mathscr{S}^I$$

<u>which assigns to every space</u> $Y\ \varepsilon\ \mathscr{S}$ <u>and object</u> $i\ \varepsilon\ I$ <u>the space</u> $Y \times I/i\ \varepsilon\ \mathscr{S}$.

The proof is straightforward.

Another way of saying this is

3.4 Proposition. <u>For every n-simplex</u>

$$u \quad = \quad (i_0 \xleftarrow{\alpha_1} \cdots \xleftarrow{\alpha_n} i_n) \qquad \varepsilon\ I$$

there is a natural map

$$ju\colon \Delta[n]\times \underleftarrow{\mathrm{holim}}\,\underline{X} \longrightarrow \underline{X}i_0 \qquad\qquad \in \mathcal{S}$$

which, for $n \geq 1$ *is a (higher) homotopy between the maps* $(\underline{X}\alpha_1)j(d_0u)$ and $j(d_1u),\cdots,j(d_nu)$, *i.e. the diagrams*

$$\begin{array}{ccc} \Delta[n-1]\times \underleftarrow{\mathrm{holim}}\,\underline{X} & \xrightarrow{j(d_0u)} & \underline{X}i_1 \\ {\scriptstyle d^0}\downarrow & & \downarrow{\scriptstyle \alpha_1} \\ \Delta[n]\times \underleftarrow{\mathrm{holim}}\,\underline{X} & \xrightarrow{ju} & \underline{X}i_0 \end{array}$$

$$\begin{array}{ccc} \Delta[n-1]\times \underleftarrow{\mathrm{holim}}\,\underline{X} & \searrow{\scriptstyle j(d_ku)} & \\ {\scriptstyle d^k}\downarrow & & \underline{X}i_0 \\ \Delta[n]\times \underleftarrow{\mathrm{holim}}\,\underline{X} & \nearrow{\scriptstyle ju} & \end{array} \qquad\qquad 0 < k \leq n$$

commute, and $\underleftarrow{\mathrm{holim}}\,\underline{X}$ *together with these maps has the obvious universal property*.

It will be shown in 5.6 that $\underleftarrow{\mathrm{holim}}$ satisfies the *homotopy lemma*: *If* $f\colon \underline{X}\to\underline{X}' \in \mathcal{S}^I$ *is such that, for every* $i \in I$, *the map* $fi\colon \underline{X}i \to \underline{X}i' \in \mathcal{S}$ *is a homotopy equivalence between fibrant objects, then the induced map* $\underleftarrow{\mathrm{holim}}\,\underline{X} \to \underleftarrow{\mathrm{holim}}\,\underline{X}'$ *is also a homotopy equivalence*. However, if some $\underline{X}i$ are *not* fibrant, then $\underleftarrow{\mathrm{holim}}\,\underline{X}$ may have the "*wrong*" homotopy type.

In §8 we shall interpret the functor $\underleftarrow{\mathrm{holim}}$ in terms of *homotopy categories*.

We end this section with a comment on the

3.5 Relationship to the (ordinary) inverse limit. For $\underline{X} \in \mathcal{S}^I$, the maps $I/i \to *$ induce a *natural map*

$$\varprojlim \underline{X} \longrightarrow \operatorname{holim}_{\leftarrow} \underline{X} \qquad \qquad \in \mathscr{S}$$

which, in general, is <u>not</u> a weak equivalence. For instance, for fibrant connected $X \in \mathscr{S}$, the diagram

$$* \rightrightarrows X$$

has as inverse limit either $*$ or the empty space (depending on whether both maps are the same or not), while the homotopy inverse limit has the homotopy type of the loop space of X.

§4. Examples and generalizations

We give some examples and generalizations and show that for a fibrant cosimplicial space, the homotopy inverse limit has the same homotopy type as the total space.

4.1 Examples. The following are examples for which the natural map $\varprojlim \underline{X} \to \operatorname{holim}_{\leftarrow} \underline{X}$ (3.5) is a weak equivalence and in which each $\underline{X}i$ is assumed to be fibrant:

(i) I is discrete (i.e. I contains only identity maps); then the homotopy inverse limit is the cartesian product.

(ii) I contains only two objects and one map between them; then the homotopy inverse limit reduces to the usual mapping path space (i.e. the "dual" of the mapping cylinder).

(iii) I has an initial object $i_0 \in I$ (i.e. for each $i \in I$, there is exactly one map $i_0 \to i \in I$); in this case, the natural map (3.4)

$$ji_0 \colon \operatorname{holim}_{\leftarrow} \underline{X} \longrightarrow \underline{X}i_0 \qquad \in \mathcal{S}$$

is also a weak equivalence.

(iv) Every diagram in $\mathcal{S}$ of the form

$$X' \longrightarrow X \longleftarrow X''$$

in which at least one of the maps is a fibration.

(v) Every tower of fibrations

$$\cdots \longrightarrow X_n \longrightarrow \cdots \longrightarrow X_1 \longrightarrow X_0 \quad .$$

<u>4.2 Example</u>. For $X \in \mathcal{S}$, let $\underline{e}X \in \mathcal{S}^I$ be the <u>constant</u> I-diagram, i.e. $(\underline{e}X)i = X$ for each $i \in I$ and each map is the identity map of X. Using 2.3 it is easy to show that

$$\underleftarrow{\operatorname{holim}}\ \underline{e}X \approx \hom(I,X) .$$

In this case the natural map (3.5)

$$X = \underleftarrow{\lim}\ \underline{e}X \longrightarrow \underleftarrow{\operatorname{holim}}\ \underline{e}X \simeq \hom(I,X)$$

is usually <u>not</u> a weak equivalence.

<u>4.3 Example</u>. If I and J are small categories and $\underline{X} \in \mathcal{S}^{I\times J}$, then

$$\underleftarrow{\operatorname{holim}}_j\ (\underleftarrow{\operatorname{holim}}_i\ \underline{X}(i,j)) \simeq \underleftarrow{\operatorname{holim}}\ \underline{X} \simeq \underleftarrow{\operatorname{holim}}_i\ (\underleftarrow{\operatorname{holim}}_j\ \underline{X}(i,j)).$$

Another important example is the case of

<u>4.4 Cosimplicial diagrams</u>. Let Δ be the category of finite ordered sets (2.6). Then

$$\mathcal{S}^\Delta = c\mathcal{S}$$

and the results of 2.6 and Ch.X, 5.2 imply:

<u>If</u> $\underset{\sim}{X} \in c\mathcal{S}$ <u>is fibrant, then the map</u>

$$\Delta/- \longrightarrow \underset{\sim}{\Delta} \qquad \in c\mathcal{S}$$

<u>of 2.6 induces a homotopy equivalence</u>

$$\operatorname{Tot} \underset{\sim}{X} = \hom(\underset{\sim}{\Delta},\underset{\sim}{X}) \longrightarrow \hom(\Delta/-,\underset{\sim}{X}) = \underleftarrow{\operatorname{holim}}\ \underset{\sim}{X} \qquad \in \mathcal{S} .$$

4.5 Generalizations. If $\mathcal{C}$ is a category which

(i) has (ordinary) inverse limits, and

(ii) has natural "function objects" $\hom(K,X) \in \mathcal{C}$, for $K \in \mathcal{S}$ and $X \in \mathcal{C}$,

then our definition (3.2) of homotopy inverse limits can be applied, and yields, for every small category I, a functor

$$\underleftarrow{\text{holim}}: \mathcal{C}^{I} \longrightarrow \mathcal{C}.$$

In particular, $\underleftarrow{\text{holim}}$ is defined for the categories $\mathcal{S}_*$ of pointed spaces, $\mathcal{T}$ of topological spaces, and $\mathcal{T}_*$ of pointed topological spaces, with the obvious "function objects", e.g. for $X \in \mathcal{T}$ and $K \in \mathcal{S}$

$$\hom(K,X) = X^{|K|}$$

with the compact-open topology. However, nothing really new happens in $\mathcal{S}_*$, $\mathcal{T}$ or $\mathcal{T}_*$, because the action of $\underleftarrow{\text{holim}}$ in $\mathcal{S}_*$ (resp. $\mathcal{T}_*$) can be obtained from its action in $\mathcal{S}$ (resp. $\mathcal{T}$) by "remembering" the base point, while for $\underline{X} \in \mathcal{T}^{I}$

$$\text{Sin}(\underleftarrow{\text{holim}}\ \underline{X}) \approx \underleftarrow{\text{holim}}(\text{Sin}\ \underline{X}) \qquad \in \mathcal{S}.$$

It might be interesting to consider the functor $\underleftarrow{\text{holim}}$ in other closed simplicial model categories (see [Quillen (HA)]).

§5. Cosimplicial replacement of diagrams

An important tool in the study of homotopy inverse limits is the cosimplicial replacement lemma (5.2), which states, that the homotopy inverse limit of a small diagram of spaces can be considered as the total space of the cosimplicial space obtained by applying:

5.1 The cosimplicial replacement functor $\prod^*: \mathcal{S}^I \to c\mathcal{S}$. A diagram $\underline{X} \in \mathcal{S}^I$ can be considered as a kind of "local coefficient system" on the space $I \in \mathcal{S}$ and its cosimplicial replacement $\prod^*\underline{X}$ is, roughly speaking, the resulting "cosimplicial space of twisted cochains". More precisely: the cosimplicial replacement of $X \in \mathcal{S}^I$ is the cosimplicial space $\prod^*\underline{X} \in c\mathcal{S}$, which in codimension n consists of the (product) space

$$\prod^n \underline{X} = \prod_{u \in I_n} \underline{X}i_0 \quad \in \mathcal{S} \qquad \text{where } u = (i_0 \xleftarrow{\alpha_1} \cdots \xleftarrow{\alpha_n} i_n)$$

with coface and codegeneracy maps induced by the maps

$$d^0\colon \underline{X}i_1 \xrightarrow{\underline{X}\alpha_1} \underline{X}i_0 \qquad \in \mathcal{S}$$

$$d^j\colon \underline{X}i_0 \xrightarrow{id} \underline{X}i_0 \qquad \in \mathcal{S} \qquad 0 < j \leq n$$

$$s^j\colon \underline{X}i_0 \xrightarrow{id} \underline{X}i_0 \qquad \in \mathcal{S} \qquad 0 \leq j \leq n .$$

It is not hard to see that this is the same as saying that

$$\prod^n \underline{X} = \hom((I/-)_n, \underline{X}) \qquad \in \mathcal{S}$$

and that the coface and codegeneracy maps are induced by the face and degeneracy maps in the diagram of spaces $I/-$.

This second description of $\prod^* \underline{X}$ immediately implies the

5.2 Cosimplicial replacement lemma. *The functor*

$$\mathcal{S}^I \xrightarrow{\underleftarrow{\text{holim}}} \mathcal{S}$$

admits a factorization

$$\mathcal{S}^I \xrightarrow{\prod^*} c\mathcal{S} \xrightarrow{\text{Tot}} \mathcal{S}$$

A long but straightforward argument using the first description of $\prod^* \underline{X}$ implies (see Ch.X, §4).

5.3 Proposition. *Let* $f: \underline{X} \to \underline{X}' \in \mathcal{S}^I$ *be such that* $fi: \underline{X}i \to \underline{X}'i \in \mathcal{S}$ *is a fibration for every* $i \in I$. *Then* $\prod^* f: \prod^* \underline{X} \to \prod^* \underline{X}' \in c\mathcal{S}$ *is also a fibration.*

5.4 Proposition. *Let* $f: \underline{X} \to \underline{X}' \in \mathcal{S}$ *be such that, for every* $i \in I$, *the map* $fi: \underline{X}i \to \underline{X}'i \in \mathcal{S}$ *is a weak equivalence between fibrant objects. Then the map* $\prod^* f: \prod^* \underline{X} \to \prod^* \underline{X}' \in c\mathcal{S}$ *is also a weak equivalence.*

In view of Ch.X, 5.1 and 5.2 these two propositions imply the following lemmas.

5.5 Fibration lemma. *Let* $f: \underline{X} \to \underline{X}' \in \mathcal{S}^I$ *be such that* $fi: \underline{X}i \to \underline{X}'i \in \mathcal{S}$ *is a fibration for every* $i \in I$. *Then* f *induces a fibration*

$$\underleftarrow{\text{holim}}\, f: \underleftarrow{\text{holim}}\, \underline{X} \longrightarrow \underleftarrow{\text{holim}}\, \underline{X}' \qquad \in \mathcal{S}\ .$$

5.6 Homotopy lemma. *Let* $f: \underline{X} \to \underline{X}' \in \mathcal{S}^I$ *be such that for every* $i \in I$

(i) $\underline{X}i$ *and* $\underline{X}'i$ *are fibrant, and*

(ii) the map $fi: \underline{X}i \to \underline{X}'i \in \mathcal{S}$ *is a homotopy equivalence.*

Then f *induces a homotopy equivalence.*

$$\underleftarrow{\text{holim}}\, f: \underleftarrow{\text{holim}}\, \underline{X} \simeq \underleftarrow{\text{holim}}\, \underline{X}' \qquad \in \mathcal{S} .$$

5.7 A generalization. In defining the cosimplicial replacement functor we only used the fact that the category $\mathcal{S}$ was a category with products. The definition thus also applies to other such categories, and it thus makes sense to observe that:

For $\underline{X} \in \mathcal{S}_*^I$ *with each* $\underline{X}i$ *fibrant, there are natural isomorphisms*

$$\pi_t \textstyle\prod^* \underline{X} \approx \prod^* \pi_t \underline{X} \qquad t \geq 0 .$$

We end with a remark on

5.8 The cosimplicial case. For $\underset{\sim}{X} \in c\mathcal{S}$ there is a natural isomorphism

$$\underset{\sim}{X}^n \approx \hom(\underset{\sim}{\Delta}_n, \underset{\sim}{X}) \qquad \text{for } n \geq 0$$

and hence the map $\Delta/- \to \underset{\sim}{\Delta} \in c\mathcal{S}$ of 2.6 induces a *natural map*

$$\underset{\sim}{X} \longrightarrow \textstyle\prod^* \underset{\sim}{X} \in c\mathcal{S} .$$

Moreover, application of the functor Tot to this map yields the map of 4.4

$$\text{Tot}\, \underset{\sim}{X} \longrightarrow \text{Tot} \textstyle\prod^* \underset{\sim}{X} = \underleftarrow{\text{holim}}\, \underset{\sim}{X} \qquad \in \mathcal{S} .$$

§6. The functor $\varprojlim^s$ for diagrams of (abelian) groups

We will see in §7 that the functor $\operatorname{holim}_{\leftarrow}$ is closely related to the functors $\varprojlim^s$ for diagrams of (abelian) groups. In preparation for this we here

(i) show that the usual functors $\varprojlim^s$ for diagrams of abelian groups can be expressed in terms of the cosimplicial replacement functor $\prod^*$ of §5, and

(ii) use this to extend the definition of the functor $\varprojlim^1$ to diagrams of (not necessarily abelian) groups.

First we recall from [Milnor] and [Roos]:

6.1 The usual definition of $\varprojlim^s$ for diagrams of abelian groups. Let I be a small category, let $\mathcal{A}$ be the category of abelian groups and let $\mathcal{A}^I$ be the category of I-diagrams in $\mathcal{A}$. For $i \in I$ and an injective $K \in \mathcal{A}$, there is an injective $K_i \in \mathcal{A}^I$ characterized by

$$\operatorname{Hom}_{\mathcal{A}^I}(\underline{A},K_i) \approx \operatorname{Hom}_{\mathcal{A}}(\underline{A}_i,K) \qquad \text{for all } \underline{A} \in \mathcal{A}^I .$$

Taking products of these injectives one gets "enough" injectives in $\mathcal{A}^I$ and defines the functors

$$\varprojlim^s: \mathcal{A}^I \longrightarrow \mathcal{A} \qquad s \geq 0$$

as the s-th right derived functors, in the sense of [Cartan-Eilenberg], of the inverse limit functor

$$\varprojlim: \mathcal{A}^I \longrightarrow \mathcal{A} .$$

Since $\underleftarrow{\lim}$ is left exact it follows that

<u>(i)</u> $\underleftarrow{\lim}^0 = \underleftarrow{\lim}$, <u>and</u>

<u>(ii) a short exact sequence</u>

$$* \longrightarrow \underline{A}' \longrightarrow \underline{A} \longrightarrow \underline{A}'' \longrightarrow * \qquad \varepsilon\, \mathcal{a}^I$$

<u>gives rise to a long exact sequence</u>

$$* \longrightarrow \underleftarrow{\lim}\ \underline{A}' \longrightarrow \underleftarrow{\lim}\ \underline{A} \longrightarrow \underleftarrow{\lim}\ \underline{A}'' \longrightarrow \underleftarrow{\lim}^1\ \underline{A}' \longrightarrow \underleftarrow{\lim}^1\ \underline{A} \longrightarrow \cdots \ .$$

Using the notation of 5.7 and Ch.X, 7.1 we will prove

<u>6.2 Proposition</u>. <u>Let</u> $\underline{A} \ \varepsilon\, \mathcal{a}^I$. <u>Then there are natural isomorphisms</u>

$$\underleftarrow{\lim}^s\ \underline{A} \ \approx\ \pi^s \prod{}^* \underline{A}\ \varepsilon\, \mathcal{a} \qquad \underline{\text{for}} \quad s \geq 0.$$

<u>6.3 Example</u>. For a group G and a G-module M, let $\underline{M}\ \varepsilon\, \mathcal{a}^I$ be the associated diagram, where I is the single-object category corresponding to G. Then the underlying space of I is K(G,1) and hence

$$\underleftarrow{\lim}^s\ \underline{M} \ \approx\ H^s(G;M).$$

<u>6.4 Remark</u>. Considerable work has been done on the <u>vanishing</u> of $\underleftarrow{\lim}^s$ for certain directed sets of abelian groups [Jensen], [Mitchell]. A best possible result is [Mitchell, p. 6]:

If I is the category of a partially ordered set of cardinality $\leq \aleph_k$, and $\underline{A}\ \varepsilon\, \mathcal{a}^I$, then $\underleftarrow{\lim}^s\ \underline{A} = 0$ for $s > k+1$.

This is clearly false for general I (see 6.3).

Proposition 6.2 suggests the following definition of

6.5 The functor $\varprojlim^1$ for diagrams of (not necessarily abelian) groups. Let I be a small category and let $\mathcal{G}$ be the category of groups. Then, for $\underline{G} \in \mathcal{G}^I$, we define (see Ch.X, 7.1)

$$\varprojlim^0 \underline{G} = \pi^0 \prod^* \underline{G} \qquad \in \mathcal{G}$$

$$\varprojlim^1 \underline{G} = \pi^1 \prod^* \underline{G} \qquad \in \text{(pointed sets)}.$$

It is not hard to verify that these functors have the properties

(i) $\varprojlim^0 = \varprojlim$, and

(ii) a short exact sequence

$$* \longrightarrow \underline{G}' \longrightarrow \underline{G} \longrightarrow G'' \longrightarrow * \qquad \in \mathcal{G}^I$$

gives rise to a natural exact sequence

$$* \longrightarrow \varprojlim \underline{G}' \longrightarrow \varprojlim \underline{G} \longrightarrow \varprojlim \underline{G}'' \longrightarrow \varprojlim^1 \underline{G}' \longrightarrow \varprojlim^1 \underline{G} \longrightarrow \varprojlim^1 \underline{G}''.$$

Moreover, a straightforward calculation yields that, for towers of groups, this definition of $\varprojlim^1$ agrees with the one of Ch.IX, §2.

Proof of 6.2. For I corresponding to a directed set this was proved in [Roos]. The general case requires a different approach, which is implicit in [André].

Let J be any function which assigns an abelian group Ji to each object $i \in I$, and let $S_J\colon I \to \mathcal{A}$ be the functor given by

$$S_J i_0 = \prod_{i_0 \to i} Ji$$

where the product runs over all maps in I with domain i_0. Among the S_J there are "enough" injectives for $\mathcal{A}^I$ (namely those indicated in 6.1), and the dual of [André, p. 8-13] shows that each $S_J \in \mathcal{A}^I$ satisfies

$$\pi^s \prod{}^* S_J = 0 \qquad \text{for } s > 0.$$

Using this result together with 6.1(i) and 6.1(ii) one now readily establishes the desired result.

§7. A spectral sequence for homotopy groups of homotopy inverse limit spaces

Using the cosimplicial replacement lemma 5.2 we construct, for each small diagram $\underline{X}$ of pointed fibrant spaces, a spectral sequence $\{E_r\underline{X}\}$, which is usually closely related to $\pi_*\operatorname{holim}_{\leftarrow} \underline{X}$. For cosimplicial diagrams this spectral sequence coincides, from E_2 on, with the usual one, i.e. the one of Ch.X, §6.

7.1 The spectral sequence. For a small category I and a diagram $\underline{X} \in \mathcal{S}_*^I$ such that $\underline{X}i$ is fibrant for every $i \in I$, we define the spectral sequence $\{E_r\underline{X}\}$ by (Ch.X, 6.1)

$$\{E_r\underline{X}\} = \{E_r \prod{}^* \underline{X}\} \qquad r \geq 1$$

and get, as an immediate consequence of 5.7, 6.2 and Ch.X, 7.2 that

$$E_2^{s,t}\underline{X} \approx \lim_{\leftarrow}{}^s \pi_t \underline{X} \qquad \text{for } 0 \leq s \leq t.$$

Moreover, in view of 5.3, the spectral sequence $\{E_r\underline{X}\}$ is closely related to the groups $\pi_j \operatorname{holim}_{\leftarrow} \underline{X}$, in the sense of Ch.IX, 5.4.

From this one readily deduces the following two propositions

7.2 A homotopy theoretic interpretation of $\lim\limits_{\leftarrow}{}^s$. For $\underline{A} \in \mathcal{A}^I$, denote by $K(\underline{A},n) \in \mathcal{S}^I$ the corresponding diagram of Eilenberg-MacLane spaces [May, p. 98]. Then one has

(i) For $\underline{A} \in \mathcal{A}^I$, there are natural isomorphisms

$$\pi_i \operatorname{holim}_{\leftarrow} K(\underline{A},n) \approx \lim_{\leftarrow}{}^{n-i} \underline{A} \qquad \text{for } 0 \leq i \leq n$$

$$= * \qquad \text{for } i > n.$$

(ii) For $\underline{G} \in \mathcal{G}^I$, there are natural isomorphisms

$$\pi_i \operatorname{holim}_{\leftarrow} K(\underline{G},1) \approx \lim_{\leftarrow}{}^{1-i} \underline{G} \qquad \text{for } i = 0, 1$$

$$= * \qquad \text{for } i > 1.$$

7.3 The functors $\lim\limits_{\leftarrow}{}^s$ for cosimplicial diagrams.

(i) Let $\tilde{A}$ be a cosimplicial abelian group. Then there are natural isomorphisms

$$\pi^s \tilde{A} \approx \lim_{\leftarrow}{}^s \tilde{A} \qquad \text{for } s \geq 0$$

which are induced by the natural cosimplicial maps (4.4 and 5.8)

$$\tilde{A} \longrightarrow \prod{}^* \tilde{A} \qquad \in c\mathcal{S} \; .$$

(ii) Let $\tilde{G}$ be a cosimplicial group. Then there are, similarly, natural isomorphisms

$$\pi^s \tilde{G} \approx \lim_{\leftarrow}{}^s \tilde{G} \qquad \text{for } s = 0, 1.$$

We next consider two special cases

7.4 Towers of fibrations. It is not hard to see that, for a tower of fibrations $\underline{X}$ in $\mathcal{S}_*$, the spectral sequence $\{E_r \underline{X}\}$ reduces to the short exact sequences of Ch.IX, 3.1

$$* \longrightarrow \lim_{\leftarrow}{}^1 \pi_{i+1} \underline{X} \longrightarrow \pi_i \operatorname{holim}_{\leftarrow} \underline{X} \longrightarrow \lim_{\leftarrow} \pi_i \underline{X} \longrightarrow * \; .$$

7.5 Cosimplicial spaces. If $\tilde{X} \in c\mathcal{S}_*$ is such that $\tilde{X}^n \in \mathcal{S}$ is fibrant for all $n \geq 0$, then the map $\tilde{X} \to \prod{}^* \tilde{X} \in c\mathcal{S}$ of 5.8 induces a map of spectral sequences

$$\{E_r \underset{\sim}{X}\} \longrightarrow \{E_r \prod^{*} \underset{\sim}{X}\} \qquad r \geq 1$$

and it is not hard to prove, using 7.3, that <u>this spectral sequence map is an isomorphism, from</u> E_2 <u>on</u>.

We end with some

<u>7.6 Generalizations</u>. Let $\underline{X} \in \mathcal{S}_*^I$ be such that $\underline{X}i \in \mathcal{S}$ is fibrant for every $i \in I$, and let $Y \in \mathcal{S}_*$. Then clearly

$$\operatorname{holim}_{\leftarrow} \hom_*(Y, \underline{X}) = \hom_*(Y, \operatorname{holim}_{\leftarrow} \underline{X})$$

where $\hom_*$ denotes the pointed function space (Ch.VIII, §4) and hence <u>there is a spectral sequence</u>

$$\{E_r(Y, \underline{X})\} = \{E_r \hom_*(Y, \underline{X})\} \qquad r \geq 1$$

<u>with</u>

$$E_2^{s,t}(Y, \underline{X}) = \lim_{\leftarrow}{}^s \pi_t \hom_*(Y, \underline{X})$$

<u>which is closely related (see Ch.IX, 5.4) to</u>

$$\pi_* \hom_*(Y, \operatorname{holim}_{\leftarrow} \underline{X}) = \pi_* \operatorname{holim}_{\leftarrow} \hom_*(Y, \underline{X}).$$

More generally, let J be another small category and let $\underline{Y} \in \mathcal{S}_*^J$. Then

$$\hom_*(\underline{Y}, \underline{X}) \in \mathcal{S}^{J^* \times I}$$

where J^* denotes the dual of J. Hence <u>there is a spectral sequence</u>

$$\{E_r(\underline{Y},\underline{X})\} = \{E_r \hom_*(\underline{Y},\underline{X})\} \qquad r \geq 1$$

<u>with</u>

$$E_2^{s,t}(\underline{Y},\underline{X}) = \underset{\leftarrow}{\lim}{}^s \pi_t \hom_*(\underline{Y},\underline{X})$$

<u>which is closely (Ch.IX, 5.4) related to</u>

$$\pi_* \operatorname{holim}_{\leftarrow} \hom_*(\underline{Y}, \operatorname{holim}_{\leftarrow} \underline{X}) = \pi_* \operatorname{holim}_{\leftarrow} \hom_*(\underline{Y},\underline{X}).$$

We will come back to this in Ch.XII, §4.

§8. Homotopy inverse limits in terms of homotopy categories

The homotopy inverse limit functor has the following interpretation in terms of homotopy categories.

Let $\mathrm{Ho}\mathcal{S}$ denote the *homotopy category of* $\mathcal{S}$, i.e. the localization of $\mathcal{S}$ with respect to the weak equivalences (Ch.VIII, 3.6) and let $\mathrm{Ho}(\mathcal{S}^I)$ be the *homotopy category of* $\mathcal{S}^I$, i.e. (Ch.VIII, 3.6) the localization of $\mathcal{S}^I$ with respect to the maps $f\colon \underline{X} \to \underline{Y} \in \mathcal{S}^I$ such that $fi\colon \underline{X}i \to \underline{Y}i \in \mathcal{S}$ is a weak equivalence for every $i \in I$. Furthermore let

$$E\colon \mathrm{Ho}\mathcal{S} \longrightarrow \mathrm{Ho}(\mathcal{S}^I)$$

be the functor which assigns to a space $X \in \mathcal{S}$ the corresponding "constant" diagram of spaces (4.2). Then one has

8.1 Proposition. *The functor* E *has as right adjoint the "total right derived functor"* (in the sense of [Quillen (HA), p.I, 4.3]) *of the functor* $\underleftarrow{\mathrm{holim}}$

$$\underline{\underline{R}}\ \underleftarrow{\mathrm{holim}}\colon \mathrm{Ho}(\mathcal{S}^I) \longrightarrow \mathrm{Ho}\mathcal{S} \quad .$$

In particular, *if* $\underline{X} \in \mathcal{S}^I$ *is such that* $\underline{X}i \in \mathcal{S}$ *is fibrant for every* $i \in I$, *then* $\underleftarrow{\mathrm{holim}}\ \underline{X} \in \mathcal{S}$ *represents* $\underline{\underline{R}}\ \underleftarrow{\mathrm{holim}}\ \underline{X}$.

8.2 Remark. Note that we did *not* consider the category $(\mathrm{Ho}\mathcal{S})^I$. The "constant" functor $\mathrm{Ho}\mathcal{S} \to (\mathrm{Ho}\mathcal{S})^I$ has, in general, *no* adjoints, i.e. limits do *not* exist in the homotopy category $\mathrm{Ho}\mathcal{S}$.

Proof of 8.1. We first prove that the category $Ho(\mathcal{S}^I)$ exists. To prove this it suffices [Quillen (HA)] to show that the category $\mathcal{S}^I$ is a closed simplicial model category with as weak equivalences the maps $f\colon \underline{X} \to \underline{Y} \in \mathcal{S}^I$ such that $fi\colon \underline{X}i \to \underline{Y}i \in \mathcal{S}$ is a weak equivalence for every $i \in I$. We define fibrations in $\mathcal{S}^I$ as maps $f\colon \underline{X} \to \underline{X}' \in \mathcal{S}^I$ such that $fi\colon \underline{X}i \to \underline{X}'i \in \mathcal{S}$ is a fibration for every $i \in I$, and cofibrations as maps which have the left lifting property (Ch.VIII, 3.4) with respect to maps which are both fibrations and weak equivalences in $\mathcal{S}^I$, and we consider the simplicial structure on $\mathcal{S}^I$ which comes from viewing objects in $\mathcal{S}^I$ as simplicial objects over the category $(\underline{\text{sets}})^I$ of I-diagrams of sets. The desired result then follows from [Quillen (HA), II, §4,Th.4], since $(\underline{\text{sets}})^I$ is closed under arbitrary limits and has a set $\{\underline{P}^i\}_{i \in I}$ of small projective generators, where each $\underline{P}^i \in (\underline{\text{sets}})^I$ is characterized by the natural isomorphism

$$\text{Hom}_{(\underline{\text{sets}})^I}(\underline{P}^i,\underline{Y}) \approx \underline{Y}i \qquad \text{for all } \underline{Y} \in (\underline{\text{sets}})^I .$$

Next we observe that it is not hard to verify that the object $I/- \in \mathcal{S}^I$ is cofibrant, and hence that the pair of adjoint functors of 3.3

$$- \times I/-\colon \mathcal{S} \longrightarrow \mathcal{S}^I$$

$$\underleftarrow{\text{holim}}\colon \mathcal{S}^I \longrightarrow \mathcal{S}$$

satisfy the conditions of [Quillen (HA), p.I, 4.5, Th.3]. Thus the total left derived functor

$$\underline{\underline{L}}(- \times I/-)\colon Ho\mathcal{S} \longrightarrow Ho(\mathcal{S}^I)$$

is left adjoint to the <u>total right derived functor</u>

$$\underline{\underline{R}}\ \underleftarrow{\text{holim}}: \operatorname{Ho}(\mathcal{S}^I) \longrightarrow \operatorname{Ho}\mathcal{S}$$

and the proposition now follows from the fact that

$$E = \underline{\underline{L}}(- \times I/-): \operatorname{Ho}\mathcal{S} \longrightarrow \operatorname{Ho}(\mathcal{S}^I) .$$

§9. A cofinality theorem

A functor $f: I \to J$ between small categories induces, by composition, a functor $f^*: \mathcal{S}^J \to \mathcal{S}^I$ and hence, for every diagram of spaces $\underline{X} \in \mathcal{S}^J$ a map

$$\underleftarrow{\text{holim}}\ \underline{X} \longrightarrow \underleftarrow{\text{holim}}\ f^*\underline{X} \qquad \in \mathcal{S} \ .$$

Our main purpose here is to give, in theorem 9.2, a sufficient condition in order that this map is a homotopy equivalence. To formulate this theorem we need a notion of

9.1 Left cofinal functors. Let I be a small category, let $f: I \to M$ be a (covariant) functor and, for every object $m \in M$, let f/m denote the category of which an object is any pair (i,μ) where $i \in I$ and $\mu: fi \to m \in M$, and of which a map $(i,\mu) \to (i',\mu')$ is any map $\alpha: i \to i' \in I$ which makes the following diagram commute

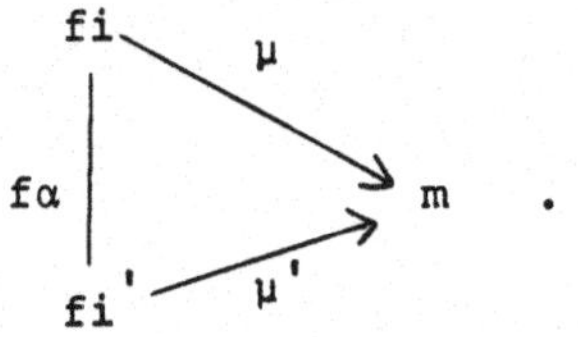

We will then say that f is *left cofinal* if, for every $m \in M$, *the space f/m is contractible*, i.e. the map $f/m \to * \in \mathcal{S}$ is a weak equivalence.

An obvious example of a left cofinal functor is the *identity functor $id: I \to I$* (see §2). Other examples will be discussed in 9.3 and 10.3.

We now state our

9.2 Cofinality theorem. *Let I and J be small categories, let $f\colon I \to J$ be a left cofinal functor, and let $\underline{X} \in \mathcal{S}^J$ be such that $\underline{X}_j$ is fibrant for all $j \in J$. Then the induced map*

$$\underleftarrow{\operatorname{holim}}\, \underline{X} \longrightarrow \underleftarrow{\operatorname{holim}}\, f^*\underline{X} \qquad \in \mathcal{S}$$

is a homotopy equivalence.

Before proving this, we show that the above notion of left cofinality agrees with the one of [Artin-Mazur, p.149] in the case where theirs was defined. More precisely:

9.3 Proposition. *Let I be a small category, which is a "left filtering"*, i.e. I is non-empty, and

(i) every pair of objects $i, i' \in I$ can be embedded in a diagram

$$\begin{array}{ccc} & & i' \\ & \nearrow & \\ i'' & & \\ & \searrow & \\ & & i \end{array}$$

(ii) if $i' \rightrightarrows i$ is a pair of maps in I, then there is a map $i'' \to i' \in I$ such that the compositions $i'' \to i$ are equal.

Then a functor $f\colon I \to M$ is left cofinal in the sense of 9.1 if and only if it is left cofinal in the sense of [Artin-Mazur], i.e. if

(iii) for every $m \in M$, there is an $i \in I$ and a map $fi \to m \in M$, and

(iv) if $m \in M$, $i \in I$ and $fi \rightrightarrows m$ are two maps in M, then there is a map $i' \to i \in I$ such that the compositions $fi' \to m$ are equal.

Proof. One easily shows that (iii) and (iv) hold if and only if $f/m \in \mathcal{S}$ is non-empty and connected for all $m \in M$. This proves the "only if" part.

Moreover, the conditions (iii) and (iv) imply that f/m is a left filtering for all $m \in M$, and the "if" part thus follows from

9.4 Proposition. If a small category I is a left filtering (9.3), then the underlying space $I \in \mathcal{S}$ is contractible, i.e. the map $I \to * \in \mathcal{S}$ is a weak equivalence.

Proof. Let $\{i_1,\cdots,i_s\}$ be a *finite* set of objects in I and let $\{\beta_1,\cdots,\beta_t\}$ be a *finite* set of maps between them. Then, because I is left filtering, there exist an object $i_0 \in I$ and maps $\alpha_j\colon i_0 \to i_j$ $(1 \le j \le s)$ such that the diagrams

$$\begin{array}{ccc} & \xrightarrow{\alpha_{u_k}} & i_{u_k} \\ i_0 & & \downarrow{\scriptstyle \beta_k} \\ & \xrightarrow[\alpha_{v_k}]{} & i_{v_k} \end{array} \qquad\qquad 1 \le k \le t$$

commute. Using this it now is not hard to show that, for every *finite* $K \in \mathcal{S}$ (i.e. K has only a finite number of non-degenerate simplices), every map $K \to I \in \mathcal{S}$ is homotopic to a constant map. Thus $I \in \mathcal{S}$ is contractible.

Proof of 9.2. Let $\prod^{**}(\underline{X},f) \in cc\mathcal{S}$ denote the *double* cosimplicial space given by

$$\prod{}^{n,q}(\underline{X},f) = \prod_{(u,\gamma,v)} \underline{X}j_0 \qquad\qquad \in \mathcal{S}$$

where

$$u = (i_0 \longleftarrow \cdots \longleftarrow i_n) \qquad \varepsilon \quad I_n$$

$$v = (j_0 \longleftarrow \cdots \longleftarrow j_q) \qquad \varepsilon \quad J_q$$

$$\gamma = fi_0 \longrightarrow j_q \qquad \varepsilon \quad J$$

with the obvious (see 5.1) pairs of coface and codegeneracy maps.

It is not hard to verify, that the <u>first</u> cosimplicial total space $\mathrm{Tot}^{(1)}\prod^{**}(\underline{X},f) \in c\mathcal{S}$ has the property that, in codimension q

$$\mathrm{Tot}^{(1)}\prod^{*,q}(\underline{X},f) = \prod_{v \in J_q} \hom(f/j_q, \underline{X}j_0) \qquad \varepsilon\ \mathcal{S}$$

where $v = (j_0 \leftarrow \cdots \leftarrow j_q) \in J_q$. The left cofinality of f implies that the maps $f/j_q \to * \in \mathcal{S}$ are weak equivalences. They therefore induce homotopy equivalences

$$\underline{X}j_0 \simeq \hom(*,\underline{X}j_0) \longrightarrow \hom(f/j_q,\underline{X}j_0) \qquad \varepsilon\ \mathcal{S}$$

which, in turn, induce a weak equivalence (see 5.1)

$$\prod^{*}\underline{X} \xrightarrow{\ \cdot\ } \mathrm{Tot}^{(1)}\prod^{**}(\underline{X},f) \qquad \varepsilon \quad c\mathcal{S}\ .$$

And as both these cosimplicial spaces are fibrant (see 5.3) application of the functor Tot yields (Ch.X, 5.2) a <u>homotopy equivalence</u>

$$\operatorname{holim}_{\leftarrow} \underline{X} = \mathrm{Tot} \prod^{*}\underline{X} \simeq \mathrm{Tot}\ \mathrm{Tot}^{(1)}\prod^{**}(\underline{X},f) \qquad \varepsilon\ \mathcal{S}\ .$$

It is also not hard to verify that the <u>second</u> cosimplicial total space $\mathrm{Tot}^{(2)}\prod^{**}(\underline{X},f) \in c\mathcal{S}$ has the property that, in codimension

n

$$\mathrm{Tot}^{(2)}\prod^{n,*}(\underline{X},f) = \prod_{u\,\varepsilon\,I_n} \mathrm{Tot}\prod^{*}(\underline{X}\backslash fi_0) \qquad \varepsilon\,\mathscr{S}$$

where $u = (i_0 \leftarrow \cdots \leftarrow i_n) \,\varepsilon\, I_n$ and $\underline{X}\backslash fi_0\colon J\backslash fi_0 \to \mathscr{S}$ denotes the diagram obtained from $\underline{X}\colon J \to \mathscr{S}$ by composition with the "inclusion" functor $J\backslash fi_0 \to J$ (see 2.7). As $J\backslash fi_0$ has an initial object, the obvious map

$$\underline{X}(fi_0) \longrightarrow \underleftarrow{\mathrm{holim}}(\underline{X}\backslash fi_0) = \mathrm{Tot}\prod^{*}(\underline{X}\backslash fi_0) \qquad \varepsilon\,\mathscr{S}$$

is a weak equivalence (4.1) and hence so is the induced map

$$\prod^{*}(f^{*}\underline{X}) \longrightarrow \mathrm{Tot}^{(2)}\prod^{**}(\underline{X},f) \qquad \varepsilon \quad c\mathscr{S}\ .$$

Again, both these spaces are fibrant (5.3) and application of the functor Tot yields (Ch.X, 5.2) a homotopy equivalence

$$\underleftarrow{\mathrm{holim}}\, f^{*}\underline{X} = \mathrm{Tot}\prod^{*}(f^{*}\underline{X}) \simeq \mathrm{Tot}\,\mathrm{Tot}^{(2)}\prod^{**}(\underline{X},f) \qquad \varepsilon\,\mathscr{S}\ .$$

The theorem now follows from the fact that the map $\underleftarrow{\mathrm{holim}}\,\underline{X} \to \underleftarrow{\mathrm{holim}}\, f^{*}\underline{X}$ and the two homotopy equivalences constructed above, can be combined into a commutative diagram

$$\begin{array}{ccc} \underleftarrow{\mathrm{holim}}\,\underline{X} & \longrightarrow & \underleftarrow{\mathrm{holim}}\, f^{*}\underline{X} \\ \downarrow & & \downarrow \\ \mathrm{Tot}\,\mathrm{Tot}^{(1)}\prod^{**}(\underline{X},f) & = & \mathrm{Tot}\,\mathrm{Tot}^{(2)}\prod^{**}(\underline{X},f)\ . \end{array}$$

The proof is straightforward, although rather long.

§10. Homotopy inverse limits for certain large diagrams of spaces

We use the cofinality theorem 9.2 to show that, for certain large (i.e. not necessarily small) diagrams of spaces, one can, at least up to homotopy, talk of their homotopy inverse limits. Our key example will be the Artin-Mazur-like large diagrams that can be obtained from a triple; and in particular we will show that the R-completion of Part I of these notes can, up to homotopy, be considered as a homotopy inverse limit of such an Artin-Mazur-like diagram.

We first describe the class of large diagrams for which our definition works.

10.1 Left small categories. A category M will be called left small if there exists a left cofinal functor $f\colon I \to M$ (with I small, of course).

Clearly every small category is left small.

Now we define

10.2 Homotopy inverse limits for left small diagrams of spaces. Let M be a left small category and let $\underline{X}$ be an M-diagram of spaces, i.e. $\underline{X} \in \mathcal{S}^M$. A homotopy inverse limit of $\underline{X}$ then will be a space of the form

$$\operatorname{holim}_{\leftarrow} f^*\underline{X} \qquad \text{where } f\colon I \to M \text{ is left cofinal.}$$

Clearly, if M is small, then $\operatorname{holim}_{\leftarrow} \underline{X}$ is a homotopy inverse limit of $\underline{X}$.

That this notion has "homotopy meaning" in general follows from

10.3 Proposition. *If $\underline{X}m$ is fibrant for all $m \in M$, then the homotopy type of $\underleftarrow{\text{holim}}\, f^*\underline{X}$ does not depend on f (or I).*

Proof. Let $f: I \to M$ and $g: J \to M$ be left cofinal functors and let $K \subset M$ be the full subcategory of M generated by the images of f and g. Then K is small and the restrictions $I \to K$ and $J \to K$ are left cofinal. Hence, if $h: K \to M$ is the inclusion functor, then 9.2 implies that the induced maps

$$\underleftarrow{\text{holim}}\, h^*\underline{X} \longrightarrow \underleftarrow{\text{holim}}\, f^*\underline{X} \qquad \text{and} \qquad \underleftarrow{\text{holim}}\, h^*X \longrightarrow \underleftarrow{\text{holim}}\, g^*\underline{X}$$

are homotopy equivalences.

In order to apply this machinery to the R-completion of Part I, we first consider

10.4 Diagrams associated with a triple. Let $\{T,\phi,\psi\}$ be a *triple* on a category $\mathcal{C}$, i.e. T is a functor $T: \mathcal{C} \to \mathcal{C}$ and ϕ and ψ are natural transformations $\phi: \text{Id} \to T$ and $\psi: T^2 \to T$ such that

$$(T\phi)\phi = (\phi T)\phi \qquad \psi(T\psi) = \psi(\psi T) \qquad \psi(T\phi) = \text{id} = \psi(\phi T).$$

An object $Y \in \mathcal{C}$ is said to *admit a T-structure* [Barr-Beck, p.337] if there is a map $\tau: TY \to Y \in \mathcal{C}$ such that $\tau\phi = \text{id}$ and $\tau\psi = \tau(T\tau)$.

For $X \in \mathcal{C}$, let $T\backslash X$ denote the full subcategory of $\mathcal{C}\backslash X$ having as objects the maps $X \to Y \in \mathcal{C}$ for which Y admits a T-structure. Then there is an obvious *Artin-Mazur-like diagram* (which sends $X \to Y$ to Y)

$$T\backslash X \longrightarrow \mathcal{C} \quad .$$

Furthermore let $\underset{\sim}{T}X$ be the <u>cosimplicial resolution</u> of X, i.e. the (augmented) cosimplicial object over $\mathcal{C}$ given by

$$(\underset{\sim}{T}X)^k = T^{k+1}X$$

in codimension k, and

$$((\underset{\sim}{T}X)^{k-1} \xrightarrow{d^i} (\underset{\sim}{T}X)^k) = (T^k X \xrightarrow{T^i \phi T^{k-i}} T^{k+1}X)$$

$$((\underset{\sim}{T}X)^{k+1} \xrightarrow{s^i} (\underset{\sim}{T}X)^k) = (T^{k+2}X \xrightarrow{T^i \psi T^{k-i}} T^{k+1}X)$$

as coface and codegeneracy maps. Then $\underset{\sim}{T}X$ (augmented) can be considered (2.6) as a functor

$$\underset{\sim}{T}X\colon \Delta \longrightarrow T\backslash X$$

and it is clear that $(\underset{\sim}{T}X)^*$ <u>carries the Artin-Mazur-like diagram $T\backslash X \to \mathcal{C}$ to the cosimplicial diagram</u> $\underset{\sim}{T}X \in c\mathcal{C}$.

This is useful because

<u>10.5 Proposition</u>. <u>The functor</u>

$$\underset{\sim}{T}X\colon \Delta \longrightarrow T\backslash X$$

<u>is left cofinal</u>.

<u>Proof</u>. Let $m\colon X \to Y$ be an object of $T\backslash X$. Then

$$\mathrm{Hom}_{T\backslash X}(\underset{\sim}{T}X, m) \qquad \in \mathcal{S}$$

is contractible (every T-structure on Y induces a contracting homotopy). Furthermore it is not hard to see, that its <u>simplicial replacement</u> (see Ch.XII, 5.1) satisfies

$$\coprod_* \text{Hom}_{T\backslash X}(\underset{\sim}{T}X, m) \simeq \underset{\sim}{T}X/m \qquad \in \mathcal{S}$$

and the desired result now follows from Ch.XII, 4.3 and 5.3.

Finally we can give our

10.6 Application to the R-completion. Let $\{R,\phi,\psi\}$ be the triple on the category $\mathcal{S}$ of Ch.I, §2. Then (10.2, 10.4 and 10.5)

(i) for every $X \in \mathcal{S}$, the space

$$\underset{\leftarrow}{\text{holim}}\ \underset{\sim}{R}X \qquad \in \mathcal{S}$$

is a homotopy inverse limit for the Artin-Mazur-like left small diagram of spaces $R\backslash X \to \mathcal{S}$, which sends a map $m: X \to Y$ into the space Y, and hence (4.4 and Ch.I, 4.2)

(ii) for every $X \in \mathcal{S}$, the R-completion of X

$$R_\infty X = \text{Tot}\ \underset{\sim}{R}X \qquad \in \mathcal{S}$$

has the homotopy type of the homotopy inverse limits of this Artin-Mazur-like diagram of spaces $R\backslash X \to \mathcal{S}$.

Chapter XII. Homotopy direct limits

§1. Introduction

In this chapter we discuss homotopy direct limits. Our account will be brief as many of the results in this chapter are dual to results in Chapter XI. Also, a construction similar to the homotopy direct limit was given by [Segal].

In slightly more detail:

§2 deals with the various (equivalent) descriptions of homotopy direct limits for the category $\mathcal{S}_*$ of pointed spaces, dualizing the results of Ch.XI, §3 and §8.

§3 dualizes the examples and generalizations of Ch.XI, §4. In particular, we observe that

(i) for a simplicial space (i.e. double simplicial set) the homotopy direct limit has the same homotopy type as the diagonal, and

(ii) our definition of $\underrightarrow{\text{holim}}: \mathcal{S}_*^I \to \mathcal{S}_*$ applies to many other categories, such as, for instance, the categories $\mathcal{S}$ of spaces, $\mathcal{T}$ of topological spaces and $\mathcal{T}_*$ of pointed topological spaces.

We also show that, for a directed system of spaces, the homotopy direct limit has the same homotopy type as the (ordinary) direct limit.

§4 In dealing with homotopy inverse limits we used proposition Ch.X, 5.2 on cosimplicial spaces. Instead of developing a similar result for simplicial spaces, we prove in §4 a proposition, which allows us to translate properties of homotopy inverse limits into properties of homotopy direct limits and then use it to prove observation (i) above, to show that the functor $\underset{\rightarrow}{\text{holim}}$ satisfies a homotopy lemma and to derive from the (homotopy) spectral sequence for homotopy inverse limits (Ch.XI, §7) a cohomology spectral sequence for homotopy direct limits.

§5 Here we obtain a simplicial replacement lemma, dual to the cosimplicial replacement lemma of Ch.XI, §5, and use it to construct a homology spectral sequence for homotopy direct limits and to recover the cohomology spectral sequence of §4.

Notation and terminology. In this chapter we will mainly work in the category $\mathcal{S}_*$ of pointed spaces.

We shall assume that the reader is familiar with ordinary direct limits for diagrams in various categories (see [Kan (AF)] and [Artin-Mazur, Appendix]).

§2. Homotopy direct limits

The homotopy direct limit of a diagram of pointed spaces is, roughly speaking, the space obtained by

(i) taking the union (i.e. wedge) of all the spaces in the diagram,

(ii) attaching to this, for every map $f: Y \to Y'$ in the diagram (which is not an identity), a copy of

$$\Delta[1] \ltimes Y = (\Delta[1] \times Y)/(\Delta[1] \times *)$$

by identifying one end with Y and the other end with Y' (as in the reduced mapping cylinder of f),

(iii) attaching to this, for every two maps $f: Y \to Y'$ and $g: Y' \to Y''$ in the diagram (neither of which is an identity), a copy of

$$\Delta[2] \ltimes Y = (\Delta[2] \times Y)/(\Delta[2] \times *)$$

by identifying the three sides with the reduced mapping cylinders of f, g and gf (or, if gf is an identity, collapsing the third side onto $Y = Y''$),

(iv) etc., etc., ...

A more efficient and precise definition is:

2.1 Homotopy direct limits. Let I be a small category and let $\underline{Y} \in \mathcal{S}_*^I$. The homotopy direct limit of $\underline{Y}$ then is the pointed space $\underrightarrow{\text{holim}}\, \underline{Y}$ (or $\underrightarrow{\text{holim}}_i\, \underline{Y}i$) defined by (see Ch.XI, 2.7)

$$\underrightarrow{\text{holim}}\, \underline{Y} = I\backslash - \ltimes \underline{Y} \qquad \in \mathcal{S}_*$$

i.e. $\underrightarrow{\text{holim}}\, \underline{Y}$ is given by the <u>difference cokernel</u> (i.e. direct limit) in $\mathcal{S}_*$ of the maps

$$\coprod_{i \xrightarrow{\gamma} i' \in I} I\backslash i' \ltimes \underline{Y}i \;\overset{a}{\underset{b}{\rightrightarrows}}\; \coprod_{i \in I} I\backslash i \ltimes \underline{Y}i$$

where $\ltimes$ is as above and a and b are respectively induced by

$$I\backslash i' \ltimes \underline{Y}i \xrightarrow{\underline{Y}\gamma} I\backslash i' \ltimes \underline{Y}i'$$

$$I\backslash i' \ltimes \underline{Y}i \xrightarrow{I\backslash\gamma} I\backslash i \ltimes \underline{Y}i \, .$$

One can, of course, obtain the functor $\underrightarrow{\text{holim}}$ also by using

<u>2.2 An adjoint functor approach</u>. <u>The functor</u>

$$\underrightarrow{\text{holim}}: \mathcal{S}_*^I \longrightarrow \mathcal{S}_*$$

<u>is left adjoint to the functor</u>

$$\text{hom}(I\backslash -,-): \mathcal{S}_* \longrightarrow \mathcal{S}_*^I \, .$$

Another way of saying this is

<u>2.3 Proposition</u>. <u>For every n-simplex</u>

$$u = (i_0 \xleftarrow{\alpha_1} \cdots \xleftarrow{\alpha_n} i_n) \qquad \varepsilon \quad I$$

<u>there is a natural map</u>

$$ju: \Delta[n] \ltimes \underline{Y}i_n \longrightarrow \underrightarrow{\text{holim}}\, \underline{Y} \qquad \varepsilon \; \mathcal{S}_*$$

<u>which, for</u> $n \geq 1$, <u>is a (higher) homotopy between the maps</u> $j(d_0u), \cdots, j(d_{n-1}u)$ <u>and</u> $j(d_nu)(\underline{Y}\alpha_n)$ <u>(see Ch.XI, 3.4), and</u>

$\underrightarrow{\text{holim}}\ \underline{Y}$ together with these maps has the obvious universal property

It will be shown in 4.2 that $\underrightarrow{\text{holim}}$ satisfies the homotopy lemma: If $f\colon \underline{Y} \to \underline{Y}' \in \mathcal{S}_*^I$ is such that, for every $i \in I$, the map $fi\colon \underline{Y}i \to \underline{Y}'i \in \mathcal{S}_*$ is a weak equivalence, then the induced map $\underrightarrow{\text{holim}}\ \underline{Y} \to \underrightarrow{\text{holim}}\ \underline{Y}'$ is also a weak equivalence.

This implies that, as for homotopy inverse limits (Ch.XI, §8), one has an interpretation of

2.4 Homotopy direct limits in terms of homotopy categories. The "constant" functor

$$E\colon \mathrm{Ho}\mathcal{S}_* \longrightarrow \mathrm{Ho}(\mathcal{S}_*^I)$$

has as left adjoint the "total left derived functor" (in the sense of [Quillen (HA), p.I, 4.3]) of the functor $\underrightarrow{\text{holim}}$

$$\underline{\underline{L}}\ \underrightarrow{\text{holim}}\colon \mathrm{Ho}(\mathcal{S}_*^I) \longrightarrow \mathrm{Ho}\mathcal{S}_* .$$

In particular, if $\underline{Y} \in \mathcal{S}_*^I$, then $\underrightarrow{\text{holim}}\ \underline{Y} \in \mathcal{S}_*$ represents $\underline{\underline{L}}\ \underrightarrow{\text{holim}}\ \underline{Y}$.

We end this section with a comment on the

2.5 Relationship to the (ordinary) direct limit. For $\underline{Y} \in \mathcal{S}_*^I$, the maps $I\backslash i \to *$ induce a natural map

$$\underrightarrow{\text{holim}}\ \underline{Y} \longrightarrow \underrightarrow{\lim}\ \underline{Y} \qquad \in \mathcal{S}_*$$

which, in general, is not a weak equivalence.

§3. Examples and generalizations

We start with dualizing the examples of Ch.XI, §4

3.1 Examples. In the following examples the natural map $\underrightarrow{\text{holim}}\ \underline{Y} \to \underrightarrow{\lim}\ \underline{Y}$ (2.5) is a weak equivalence:

(i) I is discrete; then the homotopy direct limit is the (pointed) union, i.e. the wedge.

(ii) I contains only two objects and one map between them; then the homotopy direct limit reduces to the usual reduced mapping cylinder.

(iii) I has a terminal object i_0; in this case the natural map (2.3)

$$ji_0 \colon \underline{Y}i_0 \longrightarrow \underrightarrow{\text{holim}}\ \underline{Y} \qquad \in \mathcal{S}_*$$

is also a weak equivalence.

(iv) Every diagram in $\mathcal{S}_*$ of the form

$$Y' \longleftarrow Y \longrightarrow Y''$$

in which at least one of the maps is a cofibration.

3.2 Example. For $Y \in \mathcal{S}_*$, let $\underline{e}Y \in \mathcal{S}_*^I$ be the constant I-diagram (Ch.XI, 4.2). Then

$$\underrightarrow{\text{holim}}\ \underline{e}Y \approx I \ltimes Y .$$

In this case the natural map (2.5)

$$I \ltimes Y \approx \underrightarrow{\text{holim}}\ \underline{e}Y \longrightarrow \underrightarrow{\lim}\ \underline{e}Y \approx Y$$

is, of course, usually not a weak equivalence.

3.3 Example. If I and J are small categories and $\underline{Y} \in \mathcal{S}_*^{I\times J}$, then

$$\underrightarrow{\text{holim}}_j(\underrightarrow{\text{holim}}_i\underline{X}(i,j)) \approx \underrightarrow{\text{holim}}\ \underline{X} \approx \underrightarrow{\text{holim}}_i(\text{holim}_j\underline{X}(i,j)).$$

3.4 Simplicial diagrams. Let $\underline{Y}$ be a pointed simplicial space, i.e. $\underline{Y} \in \mathcal{S}_*^{\Delta^*}$, where Δ^* denotes the dual of the category Δ (Ch.XI, 2.6). Then one can form the diagonal of $\underline{Y}$, i.e. the space $\text{diag}\ \underline{Y} \in \mathcal{S}_*$ given by

$$(\text{diag}\ \underline{Y})_n = \underline{Y}_{n,n} \qquad \text{for all } n$$

and notice that there is a natural isomorphism

$$\text{diag}\ \underline{Y} \approx \underset{\sim}{\Delta} \ltimes \underline{Y}.$$

Moreover, obviously

$$\Delta^*\backslash - = \Delta/-$$

and thus we can state:

The map $\Delta/- \rightarrow \underset{\sim}{\Delta} \in c\mathcal{S}$ of Ch.XI, 2.6, induces, for every $\underline{Y} \in \mathcal{S}_*^{\Delta^*}$, a weak equivalence

$$\underrightarrow{\text{holim}}\ \underline{Y} = \Delta/- \ltimes \underline{Y} \longrightarrow \underset{\sim}{\Delta} \ltimes \underline{Y} \approx \text{diag}\ \underline{Y} \qquad \in \mathcal{S}_*.$$

A proof of this will be given in 4.3.

Another important example is that of

3.5 Right filterings. Let I be a small category, which is a "right filtering", i.e. I is non-empty and

(i) every pair of objects $i, i' \in I$ can be embedded in a diagram

$$\begin{array}{ccc} i & & \\ & \searrow & \\ & & i'' \\ & \nearrow & \\ i' & & \end{array}$$

(ii) if $i \rightrightarrows i'$ is a pair of maps in I, then there is a map $i' \to i'' \in I$ such that the compositions $i \to i''$ are equal.

Then, for every $\underline{Y} \in \mathcal{S}_*^I$, *the natural map (2.5)*

$$\underset{\rightarrow}{\text{holim}}\ \underline{Y} \longrightarrow \underset{\rightarrow}{\lim}\ \underline{Y} \qquad \in \mathcal{S}_*$$

is a weak equivalence.

3.6 Corollary. For $Y \in \mathcal{S}_*$, *let* $\{Y_\alpha\}$ *denote the diagram of its finite pointed subspaces* (i.e. pointed subspaces with only a finite number of non-degenerate simplices). *Then the obvious map*

$$\underset{\rightarrow}{\text{holim}}_\alpha\ Y_\alpha \longrightarrow \underset{\rightarrow}{\lim}\ Y_\alpha = Y \qquad \in \mathcal{S}_*$$

is a weak equivalence.

Proof of 3.5. Let $\underline{Y}/i$ denote the I/i-diagram in $\mathcal{S}_*$, which assigns to an object $(i' \to i) \in I/i$, the space $\underline{Y}i'$. Then it is not hard to show, that the spaces $\underset{\rightarrow}{\text{holim}}\ Y/i$ form an I-diagram and that

$$\underset{\rightarrow}{\lim}_i (\underset{\rightarrow}{\text{holim}}\ \underline{Y}/i) \approx \underset{\rightarrow}{\text{holim}}\ \underline{Y} = \underset{\rightarrow}{\text{holim}}_i\ \underline{Y}i .$$

The desired result now follows readily from the fact that

(i) the map (2.5)

$$\underset{\rightarrow}{\text{holim}}\ \underline{Y}/i \longrightarrow \underset{\rightarrow}{\lim}\ Y/i = \underline{Y}i$$

is a weak equivalence (3.1(iii)), and

(ii) homotopy groups commute with direct limits of right filterings.

We end with a few comments on

3.7 Generalizations. One can dualize the remarks of Ch.XI, 4.5, and in particular, define $\underrightarrow{\text{holim}}$ for the categories $\mathcal{S}$ of spaces, $\mathcal{T}$ of topological spaces and $\mathcal{T}_*$ of pointed topological spaces. But again, nothing really new happens in these categories, as the action of $\underrightarrow{\text{holim}}$ in $\mathcal{S}$ (or $\mathcal{T}$) can be obtained from its action in $\mathcal{S}_*$ (or $\mathcal{T}_*$) by "adding a disjoint base point", while, for every $\underline{Y} \in \mathcal{S}_*^I$, one has

$$|\underrightarrow{\text{holim}}\, \underline{Y}| \approx \underrightarrow{\text{holim}}\, |\underline{Y}| \qquad \in \mathcal{T}_* \, .$$

Again, it might be interesting to consider the functor $\underrightarrow{\text{holim}}$ in other closed simplicial model categories (see [Quillen (HA)]).

§4. A relation between homotopy direct and inverse limits

In this section we prove a proposition (4.1) which allows us to translate properties of homotopy inverse limits into properties of homotopy direct limits, and use it

(i) to derive the homotopy lemma for $\underrightarrow{\text{holim}}$ (4.2), which we already mentioned in §2, from the homotopy lemma for $\underleftarrow{\text{holim}}$ (Ch.XI, 5.6),

(ii) to prove, that, for a simplicial space, the homotopy direct limit and the diagonal have the same homotopy type, using the dual result (Ch.XI, 4.4), and

(iii) derive from the (homotopy) spectral sequence for homotopy inverse limits (Ch.XI, 7.1 and 7.6), a cohomology spectral sequence for homotopy direct limits.

4.1 Proposition. *For* $\underline{Y} \in \mathcal{S}_*^I$ *and* $X \in \mathcal{S}_*$, *there is a natural isomorphism*

$$\hom_*(\underrightarrow{\text{holim}}\ \underline{Y}, X) \approx \underleftarrow{\text{holim}}\ \hom_*(\underline{Y}, X) \qquad \in \mathcal{S}_*$$

where $\hom_*$ *is the pointed function space functor (Ch.IX, 3.2).*

Proof. Let I^* denote the dual of the category I. The proposition then follows from the fact that

$$I\backslash - = I^*/-: I^* \longrightarrow \mathcal{S} .$$

One can use this to prove the

4.2 Homotopy lemma. *Let* $f\colon \underline{Y} \to \underline{Y}' \in \mathcal{S}_*^I$ *be such that* $fi\colon \underline{Y}i \to \underline{Y}'i \in \mathcal{S}_*$ *is a weak equivalence for every* $i \in I$. *Then* f *induces a weak equivalence*

$$\underrightarrow{\mathrm{holim}}\, f\colon \underrightarrow{\mathrm{holim}}\, \underline{Y} \longrightarrow \underrightarrow{\mathrm{holim}}\, \underline{Y}' \qquad \in \mathcal{S}_* .$$

Proof. It suffices to show that, for every *fibrant* $X \in \mathcal{S}_*$, the map f induces a homotopy equivalence

$$\hom_*(\underrightarrow{\mathrm{holim}}\, \underline{Y}', X) \simeq \hom_*(\underrightarrow{\mathrm{holim}}\, \underline{Y}, X) \qquad \in \mathcal{S}_* .$$

But this follows immediately from 4.1 and Ch.XI, 5.6.

Another application is to (see 3.4)

4.3 Simplicial spaces. *For a simplicial space (i.e. double simplicial set) the homotopy direct limit has the same homotopy type as the diagonal.*

Proof. Note that (see 3.4), for every $X \in \mathcal{S}_*$, there are obvious isomorphisms

$$\begin{aligned} \hom_*(\mathrm{diag}\, \underline{Y}, X) &\approx \hom_*(\underset{\sim}{\Delta} \ltimes \underline{Y}, X) \approx \\ &\approx \hom(\underset{\sim}{\Delta}, \hom_*(\underline{Y}, X)) \approx \mathrm{Tot}\, \hom_*(\underline{Y}, X). \end{aligned}$$

The proposition then follows by combining the argument of 4.2 with 4.1 and Ch.XI, 4.4.

Proposition 4.1 also allows us to reinterpret

4.4 The spectral sequence $E_r(\underline{Y}, \underline{X})$ of Ch.XI, 7.6. Let I and J be small categories. Then, for $\underline{Y} \in \mathcal{S}_*^J$ and $\underline{X} \in \mathcal{S}_*^I$, one has the spectral sequence of Ch.XI, 7.6

$$\{E_r(\underline{Y}, \underline{X})\} = \{E_r \hom_*(\underline{Y}, \underline{X})\} \qquad r \geq 1$$

with

$$E_2^{s,t}(\underline{Y}, \underline{X}) \approx \underleftarrow{\lim}^s \pi_t \hom_*(\underline{Y}, \underline{X}) \qquad 0 \leq s \leq t .$$

Moreover, if $\underline{X}i \in \mathcal{S}_*$ is fibrant for every $i \in I$, then, in view of 4.1, this spectral sequence is closely related (see Ch.IX, 5.4) to

$$\pi_* \hom_*(\operatorname{ho}\underrightarrow{\lim}\, \underline{Y}, \operatorname{ho}\underleftarrow{\lim}\, \underline{X}) .$$

A useful special case of this is:

4.5 A cohomology spectral sequence for homotopy direct limits. Let $\underline{Y} \in \mathcal{S}_*^I$ and let $\tilde{h}^*$ be a reduced generalized cohomology theory on $\mathcal{S}_*$ which "comes from a spectrum". Then 4.4 implies:

There is a natural spectral sequence $\{E_r(\underline{Y}; \tilde{h}^*)\}$ with

$$E_2^{s,t}(\underline{Y}, \tilde{h}^*) \approx \underleftarrow{\lim}^s \tilde{h}^{-t}\underline{Y} \qquad s \geq 0$$

which is closely related (see Ch.IX, 5.4) to

$$\tilde{h}^* \operatorname{ho}\underrightarrow{\lim}\, \underline{Y} .$$

An interesting case is obtained if, for $Y \in \mathcal{S}_*$, one applies this to the diagram $\{Y_\alpha\}$ of the finite (pointed) subspaces of Y (see 3.6 and [Adams, (AT), p.6]).

§5. Simplicial replacement of diagrams

Another tool in the study of homotopy direct limits is the simplicial replacement lemma (5.2), which states that the homotopy direct limit of a small diagram of spaces can be considered as the diagonal of a certain simplicial space. As an application of this we construct a homology spectral sequence for homotopy direct limits and recover the cohomology spectral sequence of 4.5.

We start with

5.1 The simplicial replacement functor $\coprod_*: \mathcal{S}_*^I \to \mathcal{S}_*^{\Delta^*}$. For $\underline{Y} \in \mathcal{S}_*^I$, its simplicial replacement is the simplicial space (i.e. double simplicial set) $\coprod_* \underline{Y}$ which in dimension n consists of the (pointed) union, i.e. wedge

$$\coprod_n \underline{Y} = \coprod_{u \in I_n} \underline{Y}i_n \quad \in \mathcal{S}_* \qquad \text{where } u = (i_0 \xleftarrow{\alpha_1} \cdots \xleftarrow{\alpha_n} i_n)$$

with face and degeneracy maps induced by the maps

$$d_j: \underline{Y}i_n \xrightarrow{id} \underline{Y}i_n \quad \in \mathcal{S}_* \qquad 0 \le j < n$$

$$d_n: \underline{Y}i_n \xrightarrow{\underline{Y}\alpha_n} \underline{Y}i_{n-1} \quad \in \mathcal{S}_*$$

$$s_j: \underline{Y}i_n \xrightarrow{id} \underline{Y}i_n \quad \in \mathcal{S}_* \qquad 0 \le j \le n .$$

It is not hard to see that this is the same as saying that

$$\coprod_n \underline{Y} = (I\backslash -)_n \ltimes \underline{Y} \qquad \in \mathcal{S}_*$$

where $\ltimes$ is as in §2, and that the face and degeneracy maps are induced by the face and degeneracy maps in the diagram of spaces I\-.

This second description of $\coprod_* \underline{Y}$ readily implies the

5.2 Simplicial replacement lemma. The functor

$$\mathcal{S}_*^I \xrightarrow{\underrightarrow{\operatorname{holim}}} \mathcal{S}_*$$

admits a factorization (see 3.4)

$$\mathcal{S}_*^I \xrightarrow{\coprod_*} \mathcal{S}_*^{\Delta^*} \xrightarrow{\operatorname{diag}} \mathcal{S}_* \quad .$$

Dualizing Ch.XI, 5.8 one then gets

5.3 The simplicial case. For a simplicial space $\underline{Y} \in \mathcal{S}_*^{\Delta^*}$, the map $\Delta/- \to \underset{\sim}{\Delta} \in c\mathcal{S}$ of Ch.XI, 2.6 induces a natural map

$$\coprod_* \underline{Y} \longrightarrow \underline{Y} \qquad \in \mathcal{S}_*^{\Delta^*}$$

of which the diagonal is exactly the map of 4.3

$$\underrightarrow{\operatorname{holim}}\ \underline{Y} \;=\; \operatorname{diag} \coprod_* \underline{Y} \longrightarrow \operatorname{diag} \underline{Y} \qquad \in \mathcal{S}_* \ .$$

5.4 A generalization. In defining the simplicial replacement functor we only used the fact that the category $\mathcal{S}_*$ was a category with sums (in this case wedges). The definition thus also applies to other such categories. For instance, one can, dualizing Ch.XI, 6.1 and 6.2, use the simplicial replacement functor to describe

5.5 The functors $\underrightarrow{\lim}^s$ for diagrams of abelian groups. For $A \in \mathcal{A}^I$, there are natural isomorphisms

$$\underset{\to}{\lim}^s \underline{A} \approx \pi_s \coprod_* \underline{A} \qquad \varepsilon\ \mathcal{a} \qquad\qquad s \geq 0$$

<u>where</u> $\underset{\to}{\lim}^s\colon \mathcal{a}^I \to \mathcal{a}$ <u>denotes the s-th left derived functor of the direct limit functor</u> $\underset{\to}{\lim}\colon \mathcal{a}^I \to \mathcal{a}$.

Combining this with 4.3 and 5.3 one gets

<u>5.6 The functors</u> $\underset{\to}{\lim}^s$ <u>for simplicial diagrams</u>. <u>For a simplicial abelian group</u> $\underline{A}$ <u>there are natural isomorphisms</u>

$$\underset{\to}{\lim}^s \underline{A} \approx \pi_s \underline{A} \qquad\qquad s \geq 0$$

<u>which are induced by the map of 5.3</u>

$$\coprod_* \underline{A} \longrightarrow \underline{A}$$

We now use these results to obtain, along the pattern of the spectral sequence of Ch. XI, 7.1,

<u>5.7 A homology spectral sequence for homotopy direct limits</u>. First we consider the <u>simplicial</u> case. For a <u>simplicial space</u> $\underline{Y} \in \mathcal{J}_*^{\Delta^*}$, one can form the <u>sequence of cofibrations</u>

$$* \longrightarrow \underset{\sim}{\Delta}^{[0]} \ltimes \underline{Y} \longrightarrow \cdots \longrightarrow \underset{\sim}{\Delta}^{[k]} \ltimes \underline{Y} \longrightarrow \cdots \qquad \varepsilon\ \mathcal{J}_*$$

and, applying to this a <u>reduced generalized homology theory</u> $\tilde{h}_*$ which "comes from a spectrum", one gets a <u>spectral sequence</u> $\{E^r(Y; \tilde{h}_*)\}$, which, <u>when</u> $\tilde{h}_*$ <u>is a connected theory, strongly converges to</u>

$$\tilde{h}_* \operatorname{diag} \underline{Y} = \tilde{h}_*(\underset{\sim}{\Delta} \ltimes \underline{Y}).$$

Moreover an argument dual to the one of Ch.X, 6.1 and 7.1, implies together with 5.6 that

$$E^2_{s,t}(\underline{Y};\ \tilde{h}_*) \approx \underset{\rightarrow}{\lim}{}^s\ \tilde{h}_t X \qquad s \geq 0.$$

Now let I be an arbitrary small category. Then we define, for $\underline{Y} \in \mathscr{S}^I_*$, its homology spectral sequence $\{E^r(\underline{Y};\ \tilde{h}_*)\}$ by

$$E^r(\underline{Y};\ \tilde{h}_*) = E^r(\coprod_* \underline{Y};\ \tilde{h}_*) \qquad r \geq 1.$$

When $\tilde{h}_*$ is a connected theory, then, in view of 5.2, this spectral sequence strongly converges to

$$\tilde{h}_*\ \underset{\rightarrow}{\operatorname{holim}}\ \underline{Y}$$

while 5.5 (always) implies that

$$E^2_{s,t}(\underline{Y};\ \tilde{h}_*) \approx \underset{\rightarrow}{\lim}{}^s\ \tilde{h}_t X \qquad s \geq 0.$$

Moreover it is not hard to prove, that, for simplicial spaces, this spectral sequence coincides, from E^2 on, with the one considered at the beginning of 5.7.

We end with observing that a similar process yields

5.8 A cohomology spectral sequence. If one replaces the reduced generalized homology theory $\tilde{h}_*$ by a reduced generalized cohomology theory $\tilde{h}^*$ which "comes from a spectrum", then the construction of 5.7 yields a cohomology spectral sequence $\{E_r(\underline{Y};\ \tilde{h}^*)\}$. It is, however, not hard to verify that this cohomology spectral sequence coincides with the one of 4.5.

Bibliography

J.F. Adams:
(S) The sphere considered as an H-space mod-p, Quart. J. Math. 12 (1961), 52-60
(AT) Algebraic topology in the last decade, Proc. Symp. Pure Math. AMS 22 (1971), 1-22

M. André:
Methode simpliciale en algèbre homologigue et algèbre commutative, Lecture Notes in Math. 32, Springer (1967)

M. Artin and B. Mazur:
Etale homotopy, Lecture Notes in Math. 100, Springer (1969)

M. Barr and J. Beck:
Acyclic models and triples, Proc. Conf. Cat. Algebra, Springer (1966), 336-343

M. Barratt and S. Priddy:
On the homology of non-connected monoids and their associated groups, Comm. Math. Helv. 47 (1972), 1-14

A.K. Bousfield and E.B. Curtis:
A spectral sequence for the homotopy of nice spaces, Trans. AMS 151 (1970), 457-479

A.K. Bousfield and D.M. Kan:
(HR) Homotopy with respect to a ring, Proc. Symp. Pure Math. AMS 22 (1971), 59-64
(LC) Localization and completion in homotopy theory, Bull. AMS 77 (1971), 1006-1010
(HS) The homotopy spectral sequence of a space with coefficients in a ring, Topology 11 (1972), 79-106
(SQ) A second quadrant homotopy spectral sequence, Trans. AMS (to appear)
(PP) Pairings and products in the homotopy spectral sequence, Trans. AMS (to appear)
(CR) The core of a ring, J. Pure Applied Algebra 2 (1972), 73-81

H. Cartan and S. Eilenberg:
Homological algebra, Princeton Univ. Press (1956)

J.M. Cohen:
Homotopy groups of inverse limits, Aarhus Univ. (1970)

E.B. Curtis:
(L) Lower central series of semi-simplicial complexes, Topology 2 (1963), 159-171
(H) Some relations between homotopy and homology, Ann. Math. 83 (1965), 386-413
(S) Simplicial homotopy theory, Aarhus Univ. (1967)

E. Dror:
(A) Acyclic spaces,
Topology (to appear)
(C) The pro-nilpotent completion of a space,
Proc. AMS (to appear)

S. Eilenberg and J.C. Moore:
Adjoint functors and triples,
Ill. J. Math. 9 (1965), 381-398

B.I. Gray:
Spaces of the same n-type, for all n,
Topology 5 (1966), 241-243

V.K.A.M. Gugenheim:
Semisimplicial homotopy theory,
Stud. Mod. Top., Prentice-Hall (1968), 99-133

M. Hall:
The theory of groups,
Macmillan (1959)

P. Hall:
Finiteness conditions for soluble groups,
Proc. London Math. Soc. 4 (1954), 419-436

D.K. Harrison:
Infinite abelian groups and homological methods,
Ann. Math. 69 (1956), 366-391

P. Hilton, G. Mislin and J. Roitberg:
Topological localization and nilpotent groups,
Bull. AMS (to appear)

C.U. Jensen:
On the vanishing of $\lim\limits_{\leftarrow}{}^{(i)}$,
J. Algebra 15 (1970), 151-166

D.M. Kan:
(HR) On the homotopy relation for c.s.s. maps,
Bol. Soc. Mat. Mex. (1957), 75-81
(AF) Adjoint functors,
Trans. AMS 87 (1958), 294-329
(AX) An axiomatization of the homotopy groups,
Ill. J. Math. 2 (1958), 548-566

D.M. Kan and G.W. Whitehead:
The reduced join of two spectra,
Topology 3 (1965), 239-261

A.G. Kurosh:
The theory of groups,
Chelsea (1955)

K. Lamotke:
Semisimpliziale algebraische Topologie,
Springer (1968)

M. Lazard:
Sur les groupes nilpotents et les anneaux de Lie,
Ann. Ec. Norm. Sup. 71 (1954), 101-190

S. MacLane:
Homology,
Springer (1963)

A.L. Malcev:
Nilpotent groups without torsion,
Jzv. Akad. Nauk. SSSR, Math. 13 (1949), 201-212

J.P. May:
Simplicial objects in algebraic topology,
Van Nostrand (1967)

J.W. Milnor:
On axiomatic homology theory,
Pac. J. Math. 12 (1962), 337-341

J.W. Milnor and J.C. Moore:
On the structure of Hopf algebras,
Ann. Math. 81 (1965), 211-264

M. Mimura, G. Nishida and H. Toda:
Localization of CW-complexes and its applications,
J. Math. Soc. Japan 23 (1971), 593-624

G. Mislin:
H-spaces mod-p (I),
Lecture Notes in Math. 196, Springer (1970), 5-10

B. Mitchell:
Rings with several objects,
Adv. Math. 8 (1972), 1-16

D.G. Quillen:
(SS) Spectral sequences of a double semi-simplicial group,
Topology 5 (1966), 155-157
(HA) Homotopical algebra,
Lecture Notes in Math. 43, Springer (1967)
(KS) The geometric realization of a Kan fibration is a Serre fibration,
Proc. AMS 19 (1968), 1499-1500
(RH) Rational homotopy theory,
Ann. Math. 90 (1969), 205-295
(PG) An application of simplicial profinite groups,
Comm. Math. Helv. 44 (1969), 45-60

D.L. Rector:
(AS) An unstable Adams spectral sequence,
Topology 5 (1966), 343-346
(EM) Steenrod operations in the Eilenberg-Moore spectral sequence,
Comm. Math. Helv. 45 (1970), 540-552

J.E. Roos:
Sur les foncteurs dérivés de $\varprojlim$,
C. R. Ac. Sci. Paris 252 (1961), 3702-3704

J. Rotman:
A completion functor on modules and algebras,
J. Algebra 9 (1968), 369-387

G. Segal:
Classifying spaces and spectral sequences,
Inst. H. Et. Sci. Math. 34 (1968), 105-112

J.-P. Serre:
Cohomologie Galoisienne,
Lecture Notes in Math. 5, Springer (1964)

E.H. Spanier:
Algebraic topology,
McGraw-Hill (1966)

J.R. Stallings:
Homology and central series of groups,
J. Algebra 2 (1965), 170-181

J. Stasheff:
(M) Manifolds of the homotopy type of (non-Lie) groups,
Bull. AMS 75 (1969), 998-1000
(H) H-spaces from a homotopy point of view,
Lecture Notes in Math. 161, Springer (1970)

N.E. Steenrod and D.B.A. Epstein:
Cohomology operations,
Princeton Univ. Press (1962)

A.E. Stratton:
A note on Ext-completions,
J. Algebra 17 (1971), 110-115

D. Sullivan:
Geometric topology, part I: localization, periodicity and Galois symmetry,
MIT (1970)

C.T.C. Wall:
Finiteness conditions for CW-complexes,
Ann. Math. 81 (1965), 56-69

G.W. Whitehead:
On mappings into group-like spaces,
Comm. Math. Helv. 28 (1954), 320-328

J.H.C. Whitehead:
A certain exact sequence,
Ann. Math. 52 (1950), 51-110

A. Zabrodsky:
Homotopy associativity and finite CW-complexes,
Topology 9 (1970), 121-128

Index

Errata

p. 34: In line 4 of 7.2, "natural left inverse" should be "natural right inverse."

p. 87: In line 6 of 6.1, "$\varinjlim H^*(Y_s; M) \approx H^*(X; R)$" should be "$\varinjlim H^*(Y_s; M) \approx H^*(X; M)$."

p. 195: The four lines of 8.5(iii) should be omitted. This alleged counterexample is incorrect, and the arithmetic square is actually a homotopy fibre square for any nilpotent space X as shown by Dror, Dwyer, and Kan in "An arithmetic square for virtually nilpotent spaces," *Ill. J. Math.*, **21**(1977), 242-254.